Afamatrix

Magic Square Grids

...the latest intrigue on Grid Combinatorics in Recreational Mathematics.

Maiden Edition

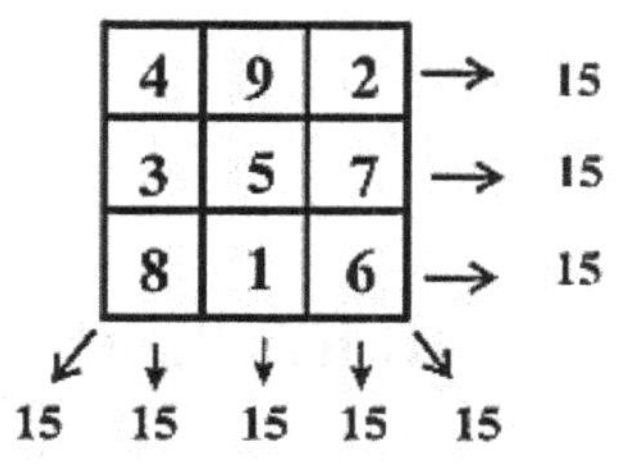

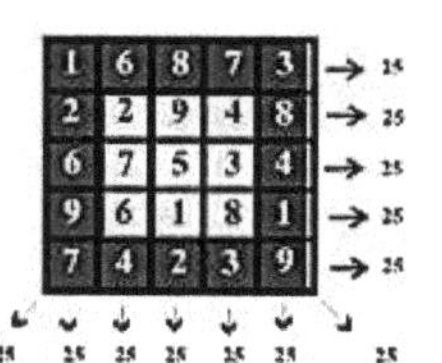

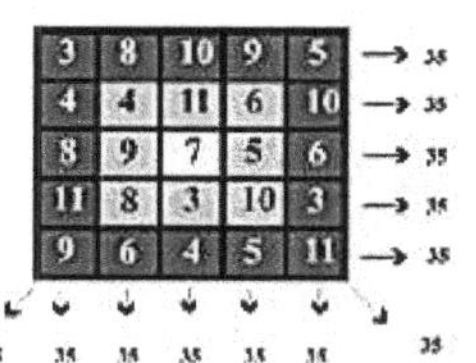

Afamefuna A. Aniagu

Intentionally left Blank

Afamatrix

Magic Square Grids

Maiden Edition

Introducing the Concept of Afamatrix Magic Square Grids, the Hyper-Afamatrix Adaptation and Afam's Method of Magic Square Grid Construction. With special features on other Magic shapes, History, Cultural Significance and the Re-classification of "Magic Squares" as "Magic Square Grids'

"For ages, men have been fascinated by Magic Squares, and since its discovery, curious thinkers has not been to rest trying to uncover the mystery and beautiful mathematical arrangement of numbers and alphabets as found in the squares, Discover yet another enigmatic and fascinating dimension of the phenomenon in Afamatrix; the latest intrigue on Grid Combinatorics in Recreational Mathematics".

By: Afamefuna A. Aniagu

Contact Information

For further enquiries, Subscription, and Distribution, Contact:

Akalusia Ent,
House 27, Road 19,
Upper-North,
Trans-Ekulu, Enugu, Nigeria

afam.aniagu@yahoo.com

+234- 8030887172

Copyright Protection

Dedication

With special reverence to the almighty God, the infinite intelligence and source of all knowledge, Jesus Christ his only son our Lord, savior of mankind and the Holy spirit, the comforter, the unseen supreme life guard;

This work is dedicated to my late mother Madam Cheluchi Victoria Aniagu whom by all possible means available inspired and encouraged me to be determined and hardworking and to always look up to the skies and say, My limit is beyond here.

Mum, you still occupy a special place in our heart, though you have been snatched away by the cold hands of death after many years of suffering and struggling to provide for us, and to be the best that mother can be to your children, all we can say is 'May your soul rest in perfect peace'. Amen

Acknowledgement

Great acknowledgement goes to my wife Ifeoma Lovina Aniagu, and my kids, whom have been by my side providing the companionship much needed for a man to soar unto a greater height. Thanks for exercising patience during the course of this work.

TABLE OF CONTENTS

CHAPTER 5

Introducing Afamatrix Magic Square Grids,

Page 28-49

CHAPTER 6

The Hyper-Afamatrix Magic Square Grid Adaptation *Page 50-58*

CHAPTER 7

Part 1 - Afam's Method of Constructing Magic Square Grids, *Page 59-155*

CHAPTER 8

Part 2 - Afam's Method of Constructing Magic Square Grids, *Page 156-191*

CHAPTER 1

Introduction

A dashed hope revived

For ages, men have been fascinated by Magic Squares, and since its discovery over a thousand years ago, Curious Thinkers has not been to rest, trying to uncover the mystery and beautiful mathematical arrangement of numbers and alphabets as found in the grids.

Discover yet another enigmatic and fascinating dimension of the phenomenon in Afam's Magic Square Grids also known as Afamatrix, The newest intrigue in the Recreational Mathematics of Combination and Permutation in Grid Combinatorics.

I started work on my **number arrangements** that I called ***Afamatrix*** without knowing that curious men of old have already done great work on the subject of Magic Squares. I thought I would be the first person to discover this wonderful and mysterious arrangement of numbers in grids. I even had prepared to turn my number arrangements into a puzzle game of almost endless Variations and Combinations, difficulty levels and sub-difficulty levels.

But one day, as I was surfing the internet in search of other types of available puzzles, for possible adaptation in the magazine I intended to publish. The magazine "***Just puzzle and fun Family Magazine***", was to focus on different Classes of puzzles, Riddles and Jokes, aimed at keeping people busy, happy and mentally fit during their leisure time. It will feature Afamatrix, and this other adapted puzzles, in other to make it more interesting (because only one type of puzzle cannot be featured alone),I stumbled into a web page in Wikipedia that displayed what looked like my Afamatrix (www.wikipedia.com/magic-square). These number arrangements were referred to as ***Magic Squares.*** I was surprised and somehow disappointed because on seeing that someone else has discovered this special type of number arrangement before me, that there is already a topic on such number arrangements, I almost lost hope on my work on Afamatrix, but fortunately for me, I did not, because I found that my brain child; Afamatrix, is a unique Magic Square of a kind, quite different from any other available Magic Square, and still makes a very interesting discovery.

The uniqueness of Afamatrix on one side is that the general formulae stated by earlier mathematicians who worked on the topic of magic squares only applies to Level 1 Afamatrix of 3 x 3 Grid only, it does not apply to any other level of Afamatrix other than that.

Secondly, I found that most of the other Classes of magic squares discovered since the history of Magic Squares had a very much different number pattern and general rules.

Afamatrix on the other hand is made up of Magic Square Grids of single digits ranging from 1 to 9 only, in various quantities on any size of grid, although figures other than single digits can be obtained in **Hyper-Afamatrix** Magic Square grids when mathematical operations like Multiplication, Division, Addition, Subtraction etc. is applied to a simple Afamatrix Magic Square Grid.

How I came about Afamatrix

I first came across this interesting arrangement of numbers in a grid during my days at Government Technical College, Enugu, Nigeria. I and my friends were trying to arrange numbers 1 to 9 in a 3 x 3 grid such that the summation of each group of three numbers will be equal to 15, The digits were drawn in a 3 x 3 grid of 9 squares altogether. We all tried but none was able solve it, We all went home with the puzzle only to come back the next day to discover that none of us had been able to arrange it. I latter tried it again and again on my own and was able to come up with a possible solution.

About ten years later, while in Germany, on a trip, I met a friend from Ukraine which I gave the 3 x 3 grid number arrangement to solve, to my surprise, I found that he already know about it. It was then I knew that the idea is universal.

Ever since then I developed interest in the subject. I tried to find out if there are other ways the numbers can be arranged, the result was positive, there are more than one possible ways of arranging the numbers, all giving a summation 15 in each row, column or diagonal, I was able to find eight possible arrangements for the 3 x 3 grid. I tried for a summation of 25 on a larger grid of 5 x 5, and I saw it was possible, with all the digits used three times except 5 which were used once. Like that, I continued for summation of 35 on a 7 x 7 grid, summation of 45 on a 9 x 9 grid and so on, they all worked. I discovered that by following a particular pattern, the summations could continue with any odd numbered grid till infinity. For instance, an equal summation of 900,000,005 per column, row and diagonals can be achieved on a sizable grid and suitable plane. The possible arrangements are only limited by available space.

On seeing the potential of these number arrangements, I decided to give it the name ***Afamatrix*** "Afam-Matriks" which is derived from the combination of the first four letters of my Igbo name; Afam-efuna and the English word; Matrix which stands for the arrangement of numbers in columns and rows and treated as a single quantity. (Advanced Learners dictionary)

The term ***Magic Square*** was added to ***Afamatrix*** after I discovered that other types of number arrangements are going by the name 'magic square'. Hence "***Afamatrix Magic Square***".

For many years I was working with odd numbered grids only, though I have in mind that it could be much possible to have even numbered Afamatrix Grids, Not until a couple of years ago that I was able to prove this to be true. As such Afamatrix of both odd numbered grids and even numbered grids can be formed.

Different variations are possible by rearranging the digits in different unique order.

Afam's Method

The Afam's Method is a new method of constructing Magic Square Grids, named after my Igbo name. This method can be used in constructing any type of Magic Square Grid. It can also be used in generating all the possible Variations of any Magic Square Grid. The Afam's method was formulated as a result of my quest to find a centralized or universal representation of Magic Square Grids, and a Construction Technique that will be Common to all Magic Square Grids.

Afamatrix Puzzle

The Afamatrix puzzle can be derived from any Afamatrix Magic Square Grid by omitting some numbers in some Squares of the grid. It can be conjugated in almost endless number of variations, combinations and difficulty levels depending on various grid sizes.

Afamatrix Code

The ***'Afamatrix Code'***, a combination of variables to represent the different Variations of Afamatrix Magic Square. It also represents the different Combinations of Afamatrix puzzle. The code can be applied to generate any Afamatrix puzzle of any variation and difficulty level.

Afamatrix Equilibrium Table

A kind of balance exists within the squares, this can be proven practically by constructing the Afamatrix equilibrium puzzle, which is a mechanical puzzle that features solid shapes of different sizes analogues to the numbers in an Afamatrix magic square. The shapes are arranged on a flat surface of a square shaped board which is rested on a pointed fulcrum. The idea is to arrange the shapes on the board such that when it is placed on the fulcrum, the board will maintain a balance.

The Idea Applied

In addition to the intellectual sharpening and mental stimulation that is derived by solving Afamatrix when applied as a puzzle, it can also be used just like other Magic squares to illustrate some fundamental concepts of arithmetic and physics, as shown in the **Afamatrix equilibrium table.** I have not been able to prove the mystical and magical properties of the squares as was indicated by ancient people that initially worked on the subject like Agrippa. Though there might be, like in *sacred geometries,* my emphasis therefore is on the physical nature of the number arrangements.

My discovery of Afamatrix Magic Square also serves to show that anyone can follow the path to original discoveries by following his or her curiosity wherever it leads, and this is a lesson that needs very much to be communicated, especially to most of we Africans-in-Africa who have almost come to the conclusion that the case of discoveries, inventions and innovations should be left to the westerners or the advanced world alone. With determination, one can make improvements and innovations.

Thus, even when very young and inexperienced, one can try to do original work in any subject. One doesn't have to wait until he/she has gone to university or specialized education, although it might take that long to figure out some of the answers, and one doesn't need anyone's permission to begin. ***That is the secret of all inventions.*** For instance, I am not a graduate when I started this work. More work is therefore still needed on Afamatrix which I have not yet been able to do because of some mathematical constraints, especially in the area of proving the equilibrium nature of Magic Squares like Afamatrix. I therefore encourage people with more mathematical prowess to come in help out, for the sake of knowledge advancement for humanity.

Can you fill in numbers in the empty boxes of the grid below such that the summation of each row, column or diagonal is equal to 105 ?

Hope you will try it before proceeding to the rest of the book? After reading this book, you will be able to do so!.

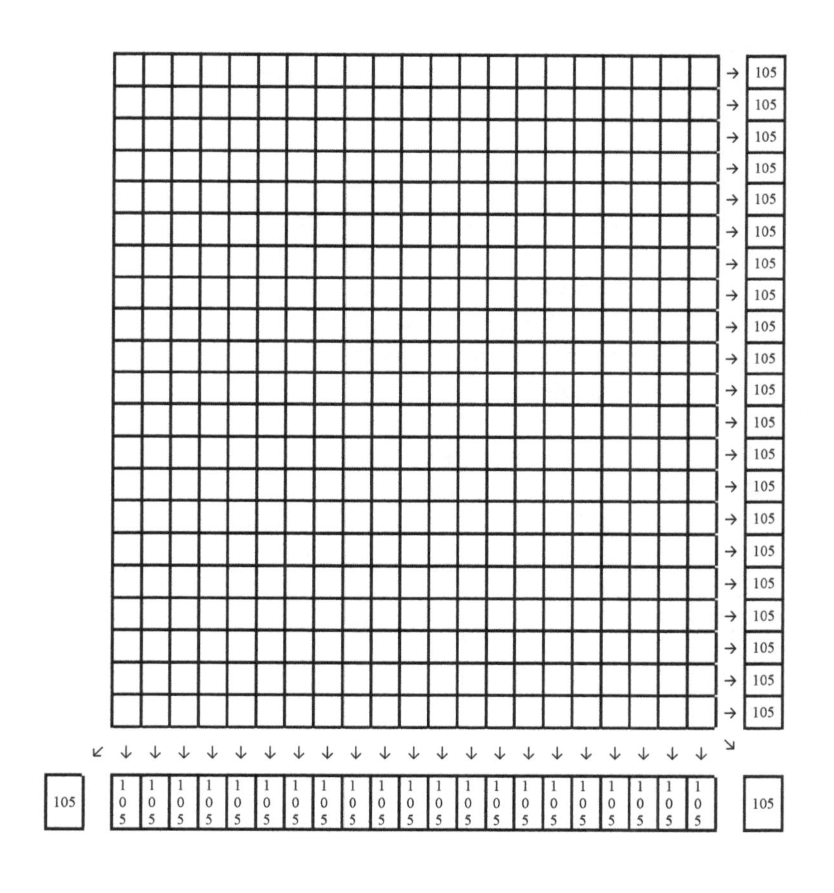
105
105
105
105
105
105
105
105
105
105
105
105
105
105
105
105
105
105
105
105
105
105
105
105

CHAPTER 2

Understanding Magic Shapes

The term "MAGIC SHAPES" had been used to refer to any of the various forms and Classes of number and word arrangements that assume a particular uniform pattern often referred to in Mathematics as Magical. However, it is penitent to understand that the "MAGIC SQUARE" term has been popularly used because of the fact that Magic Squares Grids were discovered first and such has taken precedence over others. For instance there are Magic Hexagons, Magic Heptagons, Magic Tesarracts, etc and these cannot be referred to as Magic Squares. Therefore, the right term that will give a more comprehensive Mathematical description of the subject is “MAGIC SHAPES”. This will make certain that all the magic arrangements formed with different mathematical shapes are represented.

I will also suggest the term “MAGIC SQUARE GRID” for the square arrangements instead of “MAGIC SQUARE” because the so-called Magic Squares are more of GRIDS than SQUARES. Besides, the case of one (1) number in a single square is a trivial case in magic square theory and formations while the order of (3) Magic Square, which is a grid, is the smallest non-trivial case. Also included in the Magic Square Grids are the “WORD SQUARES” which best describes the Alphabetic Or word arrangements in a square grid and should be better known as “WORD MAGIC SQUARE GRIDS”.

Therefore, the two major Categories of “MAGIC SQUARE GRIDS” should better be known as “NUMBER MAGIC SQUARE GRIDS” and “WORD MAGIC SQUARE GRIDS”, while the bridge between both will be the “NUMBER TO

WORD MAGIC SQUARE GRIDS" and "WORD TO NUMBER MAGIC SQUARE GRIDS"

Hence, both the grid (square) arrangements and the arrangements in other shapes like the cube, diamond, star etc falls under the general heading known as "MAGIC SHAPES".

As you can see from the classifications and tree diagram below, there are several other different types of magic shapes apart from the Magic Square Grids (Magic squares), Magic shapes can be classified into about five major types according to the mathematical shape with which they are formed. These are Magic Square Grid, Magic Cubes, Magic Tesarracts, Magic Stars, Magic Diamonds etc. These other magic shapes which happens to be the senior adaptations of the normal Magic Square Grid, also form a very important aspect of the magic shapes study.

On the other hand, there are two major different types of Magic Square Grids; these are the Number Magic Square Grids and the Word Magic Square Grids. There are also different types of Number Magic Square Grids, each with its own unique number arrangements; ranging from the Afamatrix Magic Square Grids, Normal Magic Square Grids, Hyper-Magic Square Grids, and Pan-Magic squares to the Bimagic and TriMagic Square Grids. There are also different types of Word Magic Square Grids, each with its own unique alphabet or word arrangements. However, Number Magic Square Grids formed the major category of Magic Square Grids. Moreover, much work had been done mainly on number Magic Square Grids where several recreational, mathematicians have been able to device different types in one form or the order.

The various types of magic shapes can be classified as shown below. We will concentrate mainly on Afamatrix Magic Square Grids, Because the details of the formations all the Magic Shapes are not within the scope of this book. More light will be thrown on the various types of Magic Shapes and its formation in subsequent publications.

Classification of Magic Shapes

1) **Magic Square Grids**
    a) **Number Magic Square Grids**
        i) **Normal Magic Square Grids**
            (a) Simple Normal Magic Square Grids
            (b) Hyper-Normal Magic Square Grids
            (c) Pan Magic Square Grids
                (i) Simple Pan Magic Square Grids
                (ii) Most Perfect Magic Square Grids
                (iii) Reversible Magic Square Grids

        ii) **Afamatrix Magic Square Grids**
            (a) Odd Order Afamatrix Magic Square Grids
                (i) Perfect Odd Order Afamatrix Magic Square Grids
                (ii) Simple Perfect Odd Order Afamatrix Magic Square Grids.
                (iii) Perfect Odd Order Hyper-Afamatrix Magic Square Grids.
                (iv) Imperfect Odd Order Afamatrix Magic Square Grids
                (v) Simple Imperfect Odd Order Afamatrix Magic Square Grids.
                (vi) Imperfect Odd Order Hyper Afamatrix Magic Square Grids.
                (vii) Perfected Imperfect Odd Order Afamatrix Magic Square Grids.eg Sudoku.

(b) Even Order Afamatrix Magic Square Grids
- (i) Perfect Even Order Afamatrix Magic Square Grids
- (ii) Simple Perfect Even Order Afamatrix Magic Square Grids
- (iii) Perfect Even Order Hyper Afamatrix Magic Square Grids
- (iv) Imperfect Even Order Afamatrix Magic Square Grids
- (v) Simple Imperfect Even Order Afamatrix Magic Square Grids.
- (vi) Imperfect Even Order Hyper-Afamatrix Magic Square Grids.
- (vii) Perfected Imperfect Even Order Afamatrix Magic Square Grids.

iii) Multi-Magic Square Grids
- (a) BiMagic Square Grids
- (b) TriMagic Square Grids

iv) Hetro Grids

(1) Simple Hetro Grids

(2) Anti-Magic Hetro Grids
- (a) Simple Anti Magic Square Grids
- (b) Complex Anti Magic Square Grids

v) Multiplicative Magic Square Grids

vi) Date Magic Square Grids

vii) Magic Square Grids of Primes

viii) Prime Reciprocal Magic Square Grids

ix) Albrecht Durers Magic Square Grids

x) The Segrada Familia Magic Square Grids

xi) Vedic Grids

xii) Sigil of Hagel Magic Square Grids

xiii) N-Queen Problem

- **xiv) Seven Magic Square Grids of orders 3 to 9 by Heinrich Cornelius Agrippa**
- **xv) Sudoku**
- **xvi) Kenken**
- **xvii) Freudenthal Magic Square Grids**
- **xviii)** Latin Grids
  - (a) Simple Latin Grids
  - (b) Graeco Latin Grids

- **xix) Number to Word Magic Square Grids**
- **xx) Word to Number Magic Square Grids**
- **xxi) Word Magic Square Grids**
  - (a) Simple Word Square Grids.
  - (b) Double Word Square Grids
- xxii) Diagonal Word Square Grids
- xxiii) Word Rectangle Grids
- xxiv) SATOR Square Grid
- xxv) Yanghui

2) Magic Cubes

a) Simple Magic Cubes

- i) Normal Magic Cubes
- ii) Perfect Magic Cubes
- iii) Diagonal Magic Cubes
- iv) Pan Diagonal Magic Cubes
- v) Perfect Magic Cubes

b) Magic Hypercubes

- i) Simple Magic Hypercubes
  - (1) Perfect Magic Hypercubes
  - (2) Semi-Perfect Magic Hypercubes
- ii) Latin Magic Hypercubes
- iii) Nasik Magic Hypercubes
- iv) Multiplicative MultimagicHypercubes

3) Magic Tessaracts

a) Perfect Magic Tessaract
b) Semi-Perfect

4) Magic Stars

a) Magic Heptagrams
b) Magic Hexagrams
c) Magic Octagrams

5) Magic Diamonds

CHAPTER 3

What is a Magic Square Grid?

Ordinarily, the question would have been "What is a Magic Square" but in the previous section, we have been able to see reasons for the use of the term "MAGIC SQUARE GRID" instead of "MAGIC SQUARE" to represent the square shaped formations.

In general '*a magic square Grid is the arrangement of numbers in rows and columns such that the summation of all the numbers in each row, column or diagonals all equals to the same constant number known as the* ***Magic constant'.***

This is the most widely accepted definition of Magic Square Grids, which this section is meant to acquaint you with. The description covers a broad range of over fifty (50) different types of Magic Square Grids and shapes which has been a result of researches from various recreational mathematicians.

The smallest Magic Square Grid is the Magic Square indeed, which is obtainable with a single square. In this case,***n***=1. (Where ***n***is is referred to as the Order of the Magic Square Grid which represents the number of Square in the Column or Row of the Magic Square Grid.) If ***n***=1, then it contains only a number inside one square, but this case is trivial, While the smallest non-trivial case is of the order 3. The case of the 2 x 2 Magic Square Grid is also trivial. As illustrated below:

1

The 1 x 1 grid magic square (A trivial case)

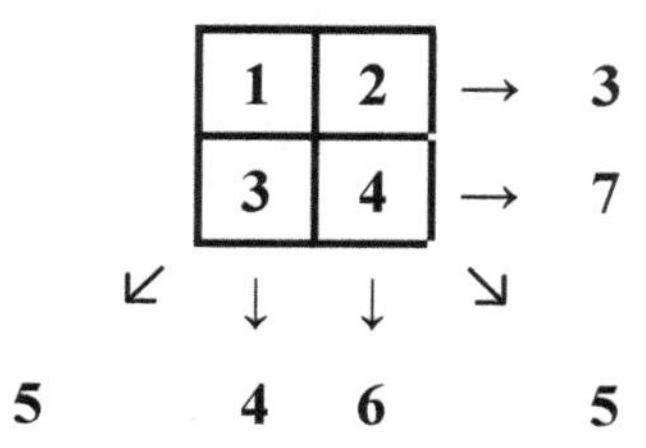

The 2 x 2 Magic Square Grid. (Also a trivial case)

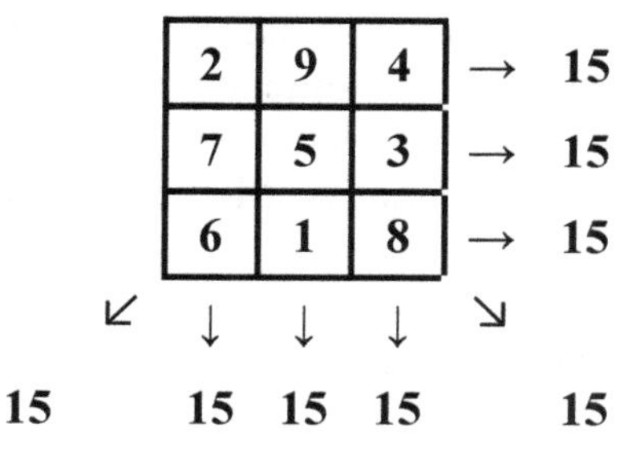

The 3 x 3 Magic Square Grid. (The Smallest non-trivial case)

Though the below definition does not apply wholly to Afamatrix Magic Square Grids as will be found in the latter chapters but there are already different mathematical expression of Magic Square Grids by different authors and researchers. However, the expression found in Wikipedia, the online encyclopedia could have given the best mathematical expression of Magic Squares so far.

Thus;

A **magic square** of order n is an arrangement of $\boldsymbol{n^2}$ numbers, usually distinct integers, in a square, such that the $\boldsymbol{n}$ numbers in all rows, all columns, and both diagonals sum to the same constant. Where ‘*n*’ represents the number of squares in each row, column or diagonals of the grid.

Substituting, Magic Square Grid for Magic Square, we will have

A **Magic Square Grid** of order n is an arrangement of n^2 numbers, usually distinct integers, in a Grid, such that the n numbers in all rows, all columns, and both diagonals sum to the same constant. Where ‘*n*’ represents the number of squares in each row, column or diagonal of the grid.

The introduction of Afamatrix has changed the conception of the Magic Square Grid phenomenon as is used to be known, this you will find out in the latter chapters where the concept and construction of Afamatrix Magic Square Grids will be treated in detail.

Shown below is an example of Order 5 or Level 2 Afamatrix Magic Square Grid.

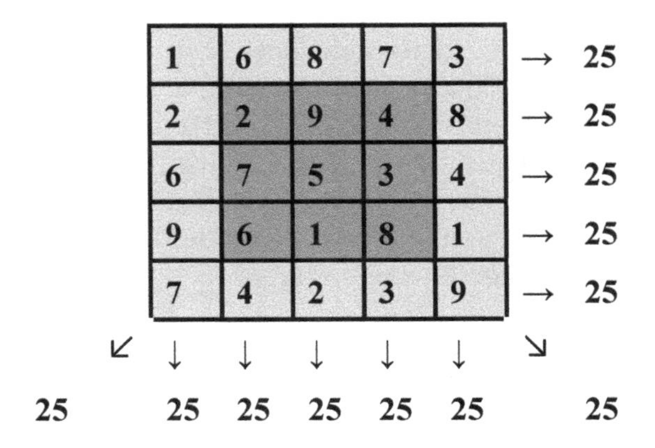

Example of a Order 5 or Level-2 Afamatrix Magic Square Grid.

General Characteristics of Magic Square Grids

The addition of any number to all the numbers in a Magic Square Grid forms a new Magic square Grid.

A new Magic Square Grid is formed if any number multiplies every number of a Magic Square Grid.

A new Magic Square Grid is formed whenever there is a change in **Frame Position** of Perfect Magic Square Grid (Refer to the appropriate chapter below on Frame Positioning for a better understanding of Magic Square Grid Frame Positions).

A new Magic Square Grid is formed whenever there is a change in Composition in any of the Horizontal or Vertical Plus Subcomponent part of any of the Frames that make up the Magic Square Grid. These changes in composition is dependent upon the various X-Frame Subcomponent Number Sequences on any of the Magic Square Grid Frames that make up the Magic Square Grid.

A new Magic Square Grid is formed if two rows, or columns, equidistant from the centre are interchanged.

A new Even Order Magic Square Grid is formed if the diagonally opposite quadrants are interchanged.

An new even order Magic Square Grid is formed if the columns or the rows are exchanged in a particular sequence. For example, in an Order-4 Magic Square Grid, If the 1st and the 2nd columns/rows are exchanged and the 3rd and the 4th columns/rows are exchanged.

CHAPTER 4

The History of Magic Square Grids

Before going into detail on the subject matter of Afamatrix and Hyper-Afamatrix Magic Square Grids, How they are constructed, the underlying Mathematical Concepts together with the Afam's method of Constructing Magic Square Grids, let's get acquainted with the history and significance of Magic Square Grids in different cultures and societies of the world.

The origin of Magic Squares (a name it has always been known for), or Magic Square Grids (a name that it should be known for) is contradictory. History has it that Magic Square Grids were found in one form or the other across different cultural and religious backgrounds. The subject has fascinated humanity for ages, and has been thought to have existed for over 4,120 years. They are found in a number of ancient civilized cultures, including China, Persia, Arabia, Northern Africa (Mali, Algeria and Egypt), Western Africa (Kanem Bornu Empire, Yoruba People of Nigeria and Senegal), India and very much latter, Europe. Let's examine the history of Magic Square Grids as was recorded across different ancient nations;

China

The oldest recorded history on Magic Square Grids is the Lo Shu square (3×3 Magic Square Grid) found in China. The principle of calculating the years and seasons using the Lo shu and Ho tu square was found in ancient Chinese calendars. Chinese literature dating from as early as 650BC tells the legend of Lo Shu or "scroll of the river Lo". The Lo Shu square has a similar origin story to the trigrams. In ancient China there was a huge flood. The great Sage, king Yu (or King Fu-hsi) who was meditating by the river Lo tried to channel the water out to sea where then emerged from the water a turtle with a curious figure/pattern on its shell; circular dots of numbers which were arranged in a three by three grid pattern such that the sum of the numbers in each row, column and diagonal was the same: 15.Fifteen also being equal to the number of days in each of the 24 cycles of the Chinese solar year. This pattern, in a certain way, was used by the people in controlling the river. Because the King was by the River Lo, this became known as the Lo Shu – the Lo River Diagram. Legend says that Yu the Great lived in the twenty-third century BC and that he quelled the Great Flood of Chinese mythology. He also set the rivers on their right courses and divided China into 9 provinces.

Lo Shu Turtle with markings.

4	9	2
3	5	7
8	1	6

The Lo Shu Square, as the Magic Square Grid on the turtle shell is called, is the unique normal Magic Square Grid of order three in which 1 is at the bottom and 2 is in the upper right corner. Every normal Magic Square Grid of order three is obtained from the Lo Shu by rotation or reflection. There is a very close relationship between Lo Shu known as the Magic Square Grid of Saturn or cronos and Ho Tu, another Magic Square Grid whose number combination has very determined relationship of the heaven and earth on the Chinese calendar, including the days and Nights. The years and seasons, can as well be deduced from this number arrangements. This same principle can also be found in other ancient calendars such as the Egyptian.

See Lo Shu – Magic Squares
(http://www.penninetaichi.co.uk/index_files/Page351.htm)

Persia

Although a definitive judgment of early history of Magic Square Grids is not available, it has been suggested that Magic Square Grids are probably of pre-Islamic Persian origin. The study of Magic Square Grids in medieval Islam in Persia is however common, and supposedly, came after the introduction of Chess in Persia. For instance in the tenth century, the Persian mathematician Buzjani has left a manuscript on page 33 of which there is a Magic Square Grids, which are filled by numbers in arithmetic progression in such a way that the sums on each line, column and diagonal are equal.

The Iron plate shown below with an order 6 Magic Square Grid in Persian/ Arabic numbers is from China and dates to the Yuan Dynasty (1271–1368).

India

In India, just like as will be found later in Egypt, Magic Square Grids are engraved on stone or metal and worn as talismans, the belief being that Magic Square Grids had astrological and divinatory qualities, their usage ensuring longevity and prevention of diseases. The 3x3 Magic Square Grid was used as part of rituals in India from vedic times, and continues to be used till date.

A well known early 4x4 Magic Square Grid in India can be seen in Khajuraho in the Parshvanath Jain temple. It dates from the 10th century. This is referred to as the ChautisaYantra, since each sub-square sums to 34.

7	**12**	**1**	**14**
2	**13**	**8**	**11**
16	**3**	**10**	**5**
9	**6**	**15**	**4**

The Kubera-Kolam is a floor painting used in India which is in the form of a Magic Square Grid of order three. It is essentially the same as the Lo Shu Square, but with 19 added to each number, giving a magic constant of 72.

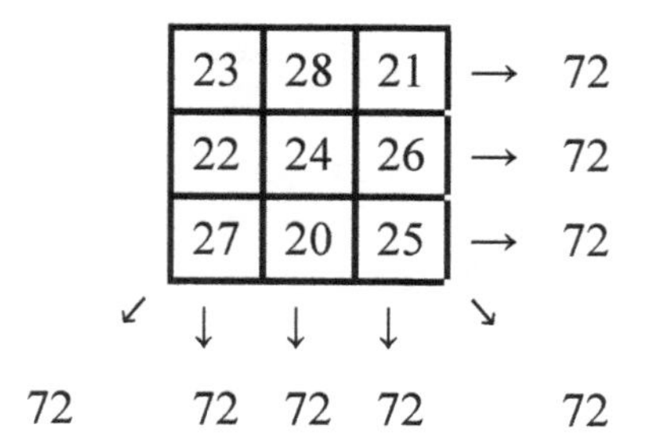

Magic Square Grids are engraved on stone or metal and worn as talismans, the belief being that Magic Square Grids had astrological and divinatory qualities, their usage ensuring longevity and prevention of diseases.

Arabia

Magic Square Grids were known to Islamic mathematicians, possibly as early as the 7th century, when the Arabs got into contact with Indians and Persians or South Asian culture, and learned Indian mathematics and astronomy, including other aspects of combinatorial mathematics. It has also been suggested that the idea came via China. The first Magic Square Grids of order 5 and 6 appear in an encyclopedia from Baghdad circa 983 AD, the Rasa'illkhwan al-Safa (the Encyclopedia of the Brethern of Purity); simpler Magic Square Grids were known to several earlier Arab mathematicians. The African Arab mathematician Ahmad al-Buni, who worked on Magic Square Grids around 1250 A.D., attributed mystical properties to them, although no details of these supposed properties are known. There are also references to the use of Magic Square Grids in astrological calculations, a practice that seems to have originated with the Arabs.

Africa

Generally, different kinds of Magic Squares have been found in various forms in North Africa, West Africa and other Sub Saharan African Nations. A Great deal of research work has been carried out on the subject of Magic Squares and it's cultural significance on the African Continent. It is believed that the idea of Magic Square migrated to Africa via Muslim traders on the old Silk Road back through the Middle East and then Northern and Western Africa, as those regions converted to Islam in the 10th century. In Africa the abstract mathematical concept combined with the long established weaving arts native to an area with an abundance of materials like cotton, silk and wool, resulted in a mesmerizing expression of design. Africans added a talismanic meaning to Magic Square Grids. Magic Square Grids, called "hatumere" were talismanic prayer papers which were sewn into clothing. Sometimes the piece of clothing itself would be adorned with symbols or phrases. The art of calligraphy and its use in decoration is a distinctive feature of Islamic design, often combined with geometric patterning on architecture, ceramics and textiles.

A collection of African artifacts and works can be found at The Textile Museum of Canada, Toronto. Credits to contemporary artists like Hamid Kachmar, Jamelie Hassan, Alia Toor and Tim Whiten. In a **2013** exhibition in Canada tagged "**Magic Squares: The patterned imagination of Muslim Africa in contemporary culture**" which showcased a lot of these work, the artists explore the relationship of patterns, communication and spirit in conversation with textiles and symbols from the Museum's permanent collection of Islamic African artifacts.Also on display is an amulet or gris-gris made of small leather pouch from Africa which contains series of Squares & Numbers based on Quranic verses. This manuscript has aMagical Square representing each surah in the Quran (total 114 Surah). The purpose of this amulet and what it cando is not reallyknown. Magic squares, known all over the world as mathematical games like Sudoku and Kenken, become carriers of powerful and diverse cultural meanings when they are painted, woven or embroidered on textiles in Muslim Africa. The conversation for the exhibition was hosted by Anthropologist and social historian ZulfikarHirji. .

Northern Africa which includes Mali, Algeria, Sudan and Egypt, has long been noted for producing skilled leatherwork, and the Magic Square pattern appears in Tuareg leather pieces from Algeria that use embellishments like incising, embossing, braiding, tussling and dying. Other embellishments, like knotting, can also have spiritual meaning, as in the hunter's coat from Mali which is on display in the textile museum of Canada. Check the website at (www.textilemuseum.ca). Old Koranic Wood Tablet Magic Squares Script has also been found in Mali. The greatest ancient library in the world is being unearthed and reassembled at Timbuktu, Mali, an astonishing archeological discovery. The discoveries showcases an amazing array of Magic Square designs engraved in stones and embroiled in fabrics and other materials. **In Egypt**, Magic Square Grids are engraved on stone or metal and worn as talismans, the belief being that Magic Square Grids had astrological and divinatory qualities, their usage ensuring longevity and prevention of diseases. Ancient Egyptian calendars adopted the principle of calculating the years and seasons using a form of Magic Square Grid known as the Lo shu and Ho tu square in China, where the 360 day year of 8640 hrs was divided by 72 to produce the 5 extra days or 120 hours on which the gods were born. The calendar shows that it takes 72 years for the heavens to move 1 degree through its Precession. Check https://www.eventbrite.ca/e/magic-squares-artists-panel-the-manifest-and-the-hidden-tickets-1360271609

In Western Africa which includes the Kanem Bornu Empire and Yoruba People of Nigeria, and Senegal, researches has shown that earlier in the 17th century, some Ulama (scholars) of Kanem -Bornu Empire, present day Northern Nigeria were highly skilled in the science of Ilm al-Awfaq (the science of Magic Squares). By the 18th century, the Borno kingdom became the most important center of learning of Mathematics in the Central Sudan attracting peoples from adjacent areas linking this at times to the occult sciences. Muhammad ibn Muhammad al-Fullani al-Kishnawi was a Fulani from northern Nigeria. In 1732 he wrote an arabic manuscript on his researches on magic squares. **The Yoruba Muslim healers from Nigeria** are known for the application of traditional healing methods which are based on Magic Squares, they are believed to have learnt there work from Ahmad al-Buni, a great Sufi mathematicial who lived in Egypt in c. 1200. Ahmad al-Buni showed how to construct magic squares using a simple bordering technique, but he may not have discovered the method himself. Al-Buni wrote about Latin squares and constructed, for example, 4 x 4 Latin squares using letters from one of the 99 names of Allah. **Senegal** is also not left out as one of the pieces on display is at the Canadian museum is a cloth that was woven to commemorate the mosque at Touba, Senegal, central to the Mouride Sufi brotherhood. The pattern features rows that alternate an image of the mosque with script that says "There is no other God but Allah", all of it over a checkerboard patterned background.

More references can be made at the buffalo.edu website at this address;http://www.math.buffalo.edu/mad/Ancient-Africa/mad_nigeria_pre-colonial.html

Europe

The Europeans came to know about Magic Square Grids when in 1300, building on the work of the Arab Al-Buni, Greek Byzantine scholar Manuel Moschopoulos wrote a mathematical treatise on the subject of Magic Square Grids, leaving out the mysticism of his predecessors. Moschopoulosis thought to be the first Westerner to have written on the subject. In the 1450s the Italian Luca Pacioli studied Magic Square Grids and collected a large number of examples. In about 1510 Heinrich Cornelius Agrippa wrote De Occulta Philosophia, drawing on the Hermetic and magical works of Marsilio Ficino and Pico della Mirandola, and in it he expounded on the magical virtues of seven magical squares of orders 3 to 9, each associated with one of the astrological planets. This book was very influential throughout Europe until the counter-reformation, and Agrippa's Magic Square Grids, sometimes called Kameas, continue to be used within modern ceremonial magic in much the same way as he first prescribed. Details of the seven magical square of Agrippa together with the work of early Europeans like the sigil of Hagel, Some word square, Albrecht Dürer's Magic Square Grid and the Sagrada Família Magic Square Grid can easily be found Magic Square Grid write ups.

Magic Square Grids Today

The study of Magic Square Grids have so much evolved in recent times, that the subject can no longer be comfortably referred to as Magic Square Grids. The term “Magic Shapes” gives a broader representation since different mathematical shapes are now involved. Different researchers, mathematicians and hobbyist alike have written on the subject of Magic Shapes which Magic Square Grids forms the major part, and such, several books, journals etc includes Magic Square Grid topics in whole or in part. Several write-ups have been published online on the internet as well. The problem facing the study of Magic Shapes today is that Most professional mathematicians consider "Magic Shapes" to be purely recreational mathematics and unworthy of their attention. Nevertheless, I believe that, partly as a result of the recent discoveries made on this topic, the attitude may change in the future. The reasons are that some mathematicians have shown that it is really possible to prove general theorems about Magic Square Grids and Shapes; secondly, the techniques of constructing Magic shapes are beginning to involve more and more sophisticated mathematical objects, the study of which is of independent interest to professional mathematicians as shown by the various publications on the topic. I am not particularly interested in the mystical properties of magic shapes, though some authors have tried to show that such properties exist. Like the 6 x 6 Magic Square Grid which uses the number 1 to 36 with a magic constant of 111, and when all the magic constants are summed will give the number 666, it might interest you to know that the number 1, 2, 3, ...36 also sum up to 666, the biblical number of the beast. I therefore will encourage many to play on the number to see the extent one can get on the discover

CHAPTER 5

Introducing Afamatrix Magic Square Grids

Having known What Magic Square Grids or Shapes is all about, together with it's History, It is high time I acquaint you with '***Afamatrix Magic Square Grids***', The main topic of this book. This book was put together, not only to make a public presentation of the Afam's Method of constructing Magic Square Grids, but, also to bring to limelight my special and very unique type of Magic Square Grid called *Afamatrix*. In the course of this chapter, we will be looking at the formation of Afamatrix Magic Square Grids and the difference between Afamatrix Magic Square Grids and other Magic Square Grids. The mathematics of Afamatrix is another important aspect that will be examined. Shown below is an example of level 2 (Order 5) Afamatrix Magic Square Grid.

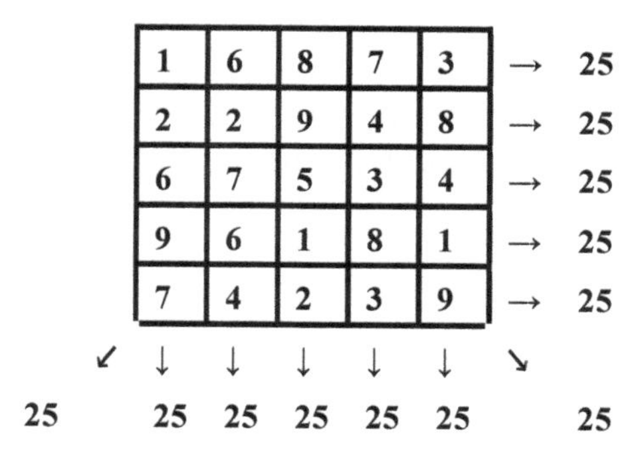

1	6	8	7	3	→ 25
2	2	9	4	8	→ 25
6	7	5	3	4	→ 25
9	6	1	8	1	→ 25
7	4	2	3	9	→ 25
↓ 25	↓ 25	↓ 25	↓ 25	↓ 25	

↙ 25   ↘ 25

Mathematically;

An **Afamatrix Magic Square Grid** of Order ***n*** is an arrangement of n^2 numbers of single digits in a square grid, such that the sum of the numbers in any of the rows, columns, or diagonals is constant and equal to ***5n***, known as the Magic

Constant ***M***, where ***n*** represents the number of squares in each row, column or diagonals of the grid.

i.e. **M=5n**

By derivation;
Afamatrix Magic Square Grids can be expressed as the category of Magic Square Grids having an arrangement of single digit numbers inside a Square Grid, such that the summation of all the numbers in any of the row, column or diagonals is constant and equal to a Magic Sum '***M***' whose change is directly proportional to the change in the Magic Square Grid Order '***n***', with the constant '***K***' of change being equal to Five (5).
Therefore, the relationship between the Magic Sum '***M***' and Order '***n***' is proportional, and given by:

$$\mathbf{M \propto n}$$

$$\mathbf{M=kn}$$

If **K=5** then **M=5n**

Figures other than single digits can be obtained in **Hyper-Afamatrix** Magic Square Grids when mathematical operations like Multiplication, Division, Addition, Subtraction etc, is applied to a simple Afamatrix Magic Square Grid.

Levels in Afamatrix Magic Square Grids

Afamatrix is can expressed in "Order" like other Magic Square Grids, just like it can also be expressed in "Levels" "L" .

The table below shows the relationship between the Levels and Order of an Afamatrix Magic Square grid.

Relationship Between Order and Level in Afamatrix Magic Square Grids

Showing the first Ten Levels / Order , Normal Magic Square Grids are said to exist for all orders of n≥ 1, except n=2.

ORDER (n)	LEVEL (L)	GRID SIZE ($n^{2)}$)	AFAMATRIXMAGIC CONSTANT (M)	REMARKS
1	0	1 * 1 = 1	5	Trivial Case
2	0.5	2 * 2 = 4	10	"
3	1	3 x 3 = 9	15	Non-Trivial Case
4	1.5	4 x 4 = 20	20	"
5	2	5 x 5 =25	25	"
6	2.5	6 x 6 = 36	30	"
7	3	7 x 7 = 49	35	"
8	3.5	8 x 8 = 64	40	"
9	4	9 x 9 = 81	45	"
10	4.5	10 x 10 = 100	50	"

$$\mathbf{L = (n-1)/2}$$

$$\mathbf{M = 10L + 5}$$

$$\mathbf{N=(2L + 1)^2}$$

Hence;

An **Afamatrix Magic Square Grid** of Level L is an arrangement of $(2L + 1)^2$ numbers of single digits in a square grid, such that the sum of the numbers in any of the rows, columns, or diagonals is constant and equal to $10L + 5$, known as the Afamatrix Magic Constant M, Where $2L + 1$ is equal to the number (n) of squares in each row, column or diagonal of the grid.

$$M = 10L + 5$$

The above definitions and formulae form the main basis of Afamatrix Magic Square Grids. Afamatrix Magic Square Grids has simple, as well as complex Mathematical expressions, This will be treated fully in the chapter titled "The mathematics of Afamatrix Magic Square Grid" however let us see how Afamatrix Magic Square Grid compares with other Magic Square Grids before going to the Mathematical calculations and derivations of Afamatrix formulae.

Comparison of Afamatrix and other Magic Square Grids.

There are at least four basic difference between Afamatrix and other Magic Square Grids. These have to do with:

1. Difference in Number Selection
2. Difference in Number of Digits
3. Difference in Magic Constant
4. Difference in Centre Square Number

1. Difference in Number Selection

According to the rule, a **normal** Magic Square Grid contains Digits from 1 to n^2 which cannot be repeated. But this applies only to a 3 x 3 Afamatrix Magic Square Grid. Afamatrix Grids other than 3 x 3 do not obey the rule. A normal Afamatrix Magic Square Grid on the other hand is made up of various sizes of grids containing single digits ranging from 0to 9 or 1 to 9, as the case may be, in various quantities according to the Grid sizes.

Figures other than single digits can be obtained in **Hyper-Afamatrix** grids when mathematical operations like Multiplication, Division, Addition, Subtraction etc., is applied to a simple Afamatrix grid.

2. Difference in Number of Digits

The number of each digit used increases in proportion to the grid size. I.e. 3 x3 grid uses 1 of each digit; 5 x 5 grid uses 3 of each digit; 7 x 7 grid uses 6 of each digit; etc, with the exception of digit 5 which is either used once or not used at all.

Generally, in Afamatrix, the increase in the number of each digit used as the grid size increases forms a special Class of arithmetic progression represented by the formulae:

$$S = \frac{n_f}{2}(2+(n_f-1) \quad = \quad S = \frac{n_f}{2}(1+L) \quad = \quad S = \frac{N-1}{8}$$

Where

S= *Sum total of each integer used,*
N = *total number of squares in the entire grid,*
L= *Level*
n_f= *Frame number*
See the section on the mathematics of Afamatrix for details on the derivation of the above formulae.

3. Difference in Magic Constant

For every Magic Square Grid, the constant sum in every row, column and diagonal is called the MAGIC CONSTANT or MAGIC SUM, denoted by *M.*

Former theories on Magic Square Grids had it initially stated that the magic constant of a normal Magic Square Grid depends only on *n* and has the value:

$$M = \frac{n^3 + n}{2} \qquad \text{OR} \qquad M = \left(\frac{n^2}{2} + 0.5\right) n$$

That is to say that for normal Magic Square Grids of order n=3, 4, 5,....., the magic constants are: 15, 34, 65, 111, 175, etc

But this also does not apply to Afamatrix Magic Square Grid because an Afamatrix Magic Square Grid of order n=3,4,5,6,7,8,9,..,..,... etc has the magic constants of 15,20,25,30,35,40,45,..,..,... etc and the relationship can be deduced from the formula: **M=5 x n**

4. **Difference in Centre Square Number**

The Centre number of a normal Magic Square Grid is given as:

$$M = \left(\frac{n^2}{2} + 0.5 \right) n$$

But this also does not apply to Afamatrix which has a constant centre square number "5" no matter the size of the Afamatrix Magic Square Grid, the middle number only changes in relation to 5 when any of the mathematical operations like, addition, multiplication,etc are applied to a simple Afamatrix Magic Square grid.

Classification of Afamatrix Magic Square Grids

As shown in the previous chapter on "Classification of Magic Shapes" , There are two main Classes of Afamatrix Magic Square Grids, the Even Order Afamatrix Magic Square Grids and the Odd Order Afamatrix Magic Square Grids, any of these two categories can be further divided into the Perfect Afamatrix Magic Square Grids, Semi-Perfect Afamatrix Magic Square Grids, Perfected Imperfect Afamatrix Magic Square Grids and the Imperfect Afamatrix Magic Square

Grids. Any of these Four (4) Categories can be further divided into Complementary or Non-Complementary Magic Square Grids depending on the arrangement of the Frame Elements. Any of the Complementary or Non-Complementary Magic Square Grids can be further classified according to the number of sections that forms the Magic Square Grid and finally, the Magic Square Grid can be classified as either Simple-Afamatrix or Hyper-Afamatrix Magic Square Grids. Hyper-Afamatrix Magic Square Grids can be obtained when arithmetic operations like Multiplication, Division, Addition, Subtraction etc. is applied to a simple Afamatrix Magic Square Grid. The picture below shows the classification model of Afamatrix Magic Square Grids.

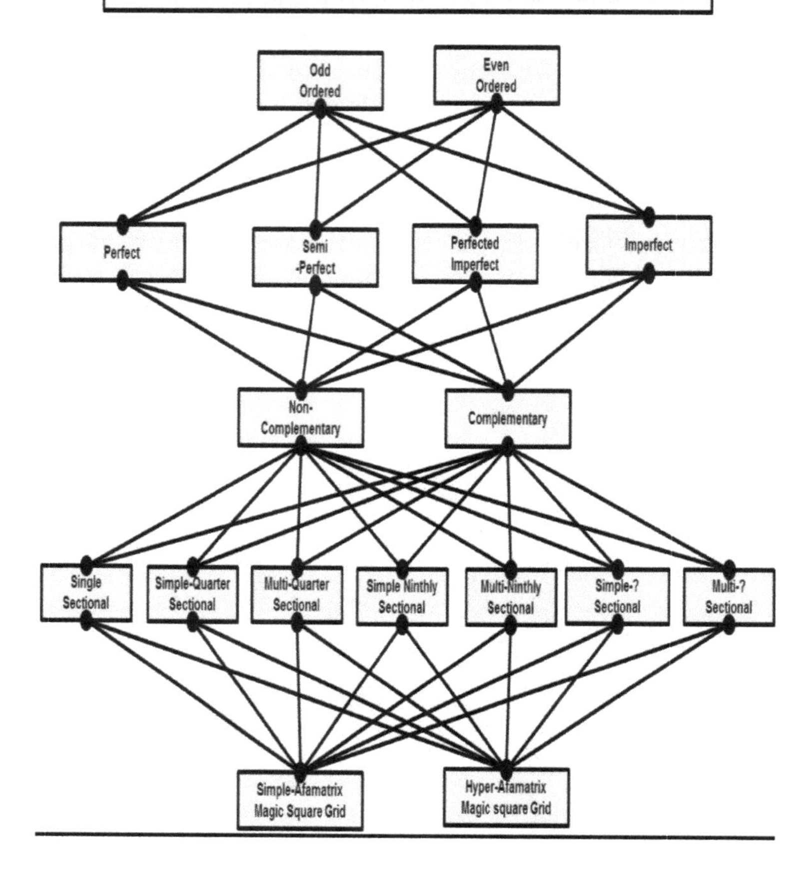
CLASSIFICATION MODEL OF AFAMATRIX MAGIC SQUARE GRIDS
Odd Ordered
Even Ordered
Perfect
Semi -Perfect
Perfected Imperfect
Imperfect
Non-Complementary
Complementary
Single Sectional
Simple-Quarter Sectional
Multi-Quarter Sectional
Simple Ninthly Sectional
Multi-Ninthly Sectional
Simple-? Sectional
Multi-? Sectional
Simple-Afamatrix Magic Square Grid
Hyper-Afamatrix Magic square Grid

1. Simple Perfect Afamatrix Magic Square Grids

All the Afamatrix Magic Square Grids that obey the complete rule of Afamatrix are classified as Simple Perfect Afamatrix Magic Square Grids.

7	4	1	8	→ 20
8	1	4	7	→ 20
3	6	9	2	→ 20
2	9	6	3	→ 20

↙ ↓ ↓ ↓ ↓ ↘

20 20 20 20 20 20

Example of a Simple Perfect Afamatrix Magic Square Grid (Note that the equal usage of all the digits)

Rules of Simple Perfect Afamatrix Magic Square Grids

The rules of a Simple Perfect Afamatrix Magic Square Grid include:

1) Only the digits 1 to 9 are used
2) The digits 1, 2, 3, 4, 6, 7, 8, 9 are used equally.

2. <u>Simple Semi-Perfect or Semi-Imperfect Afamatrix Magic Square Grids</u>

All the Afamatrix Magic Square Grids that Partially obey the rules of Afamatrix are classified as Simple Semi-Perfect Afamatrix Magic Square Grids.

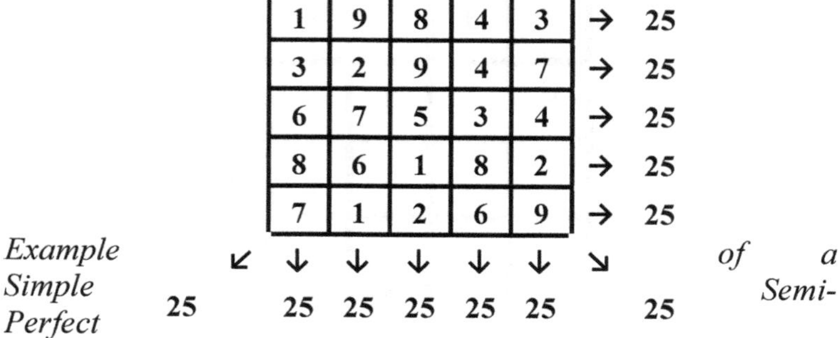

	1	9	8	4	3	→	25
	3	2	9	4	7	→	25
	6	7	5	3	4	→	25
	8	6	1	8	2	→	25
	7	1	2	6	9	→	25
↙	↓	↓	↓	↓	↓	↘	
25	25	25	25	25	25	25	

Example of a Simple Semi-Perfect Afamatrix Magic Square Grid
(Note that the equal usage of all the digits, i.e. 3 times each, except digit 5 that is used only once)

<u>Rules of Simple Semi-Perfect Afamatrix Magic Square Grids</u>

The rules of a Simple Perfect Afamatrix Magic Square Grid include:

1) Only the digits 1 to 9 are used
2) All the digits are used equally, with the exception of any of the digits which is used only once or not used at all.

3. <u>Simple Imperfect Afamatrix Magic Square Grids</u>

These comprises of all the Afamatrix Magic Square Grids that do not fully obey the complete rule of Afamatrix. They include the Afamatrix Magic Square Grids that do not contain an equal amount of each a number in the entire grid. Though Imperfect Afamatrix Magic Square Grids forms an integral and interesting part of Afamatrix, I will pay more attention to Perfect Afamatrix Magic Square Grids for our study, for the sake of convenience.

	1	6	8	7	3		
	1	6	8	7	3	→	25
	3	2	9	4	7	→	25
	6	7	5	3	4	→	25
	8	6	1	8	2	→	25
	7	4	2	3	9	→	25
↙	↓	↓	↓	↓	↓	↘	
25	25	25	25	25	25		25

Example of a Simple Imperfect Afamatrix Magic Square Grid. (Note that the digits are not used equally; digit 1 is used two times while digit 3 is used four times, the rest of the digits are used three times each)

Some Examples of Afamatrix Magic Square Grids

1. The 8 Simple Perfect Order 3 Afamatrix Magic Square Grids

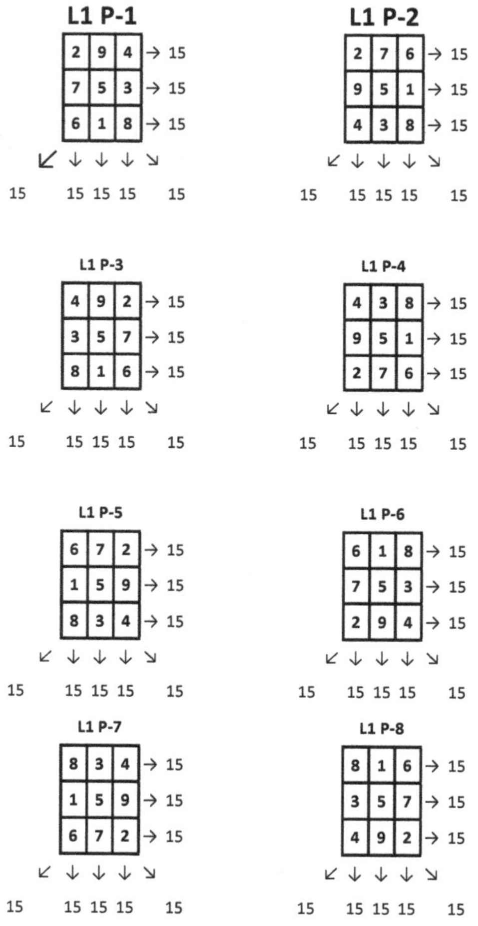

L1 P-1

2	9	4	→ 15
7	5	3	→ 15
6	1	8	→ 15

↙ ↓ ↓ ↓ ↘

15 15 15 15 15

L1 P-2

2	7	6	→ 15
9	5	1	→ 15
4	3	8	→ 15

↙ ↓ ↓ ↓ ↘

15 15 15 15 15

L1 P-3

4	9	2	→ 15
3	5	7	→ 15
8	1	6	→ 15

↙ ↓ ↓ ↓ ↘

15 15 15 15 15

L1 P-4

4	3	8	→ 15
9	5	1	→ 15
2	7	6	→ 15

↙ ↓ ↓ ↓ ↘

15 15 15 15 15

L1 P-5

6	7	2	→ 15
1	5	9	→ 15
8	3	4	→ 15

↙ ↓ ↓ ↓ ↘

15 15 15 15 15

L1 P-6

6	1	8	→ 15
7	5	3	→ 15
2	9	4	→ 15

↙ ↓ ↓ ↓ ↘

15 15 15 15 15

L1 P-7

8	3	4	→ 15
1	5	9	→ 15
6	7	2	→ 15

↙ ↓ ↓ ↓ ↘

15 15 15 15 15

L1 P-8

8	1	6	→ 15
3	5	7	→ 15
4	9	2	→ 15

↙ ↓ ↓ ↓ ↘

15 15 15 15 15

2. All the 40 Possible Order 3 Afamatrix Magic Square Grids:

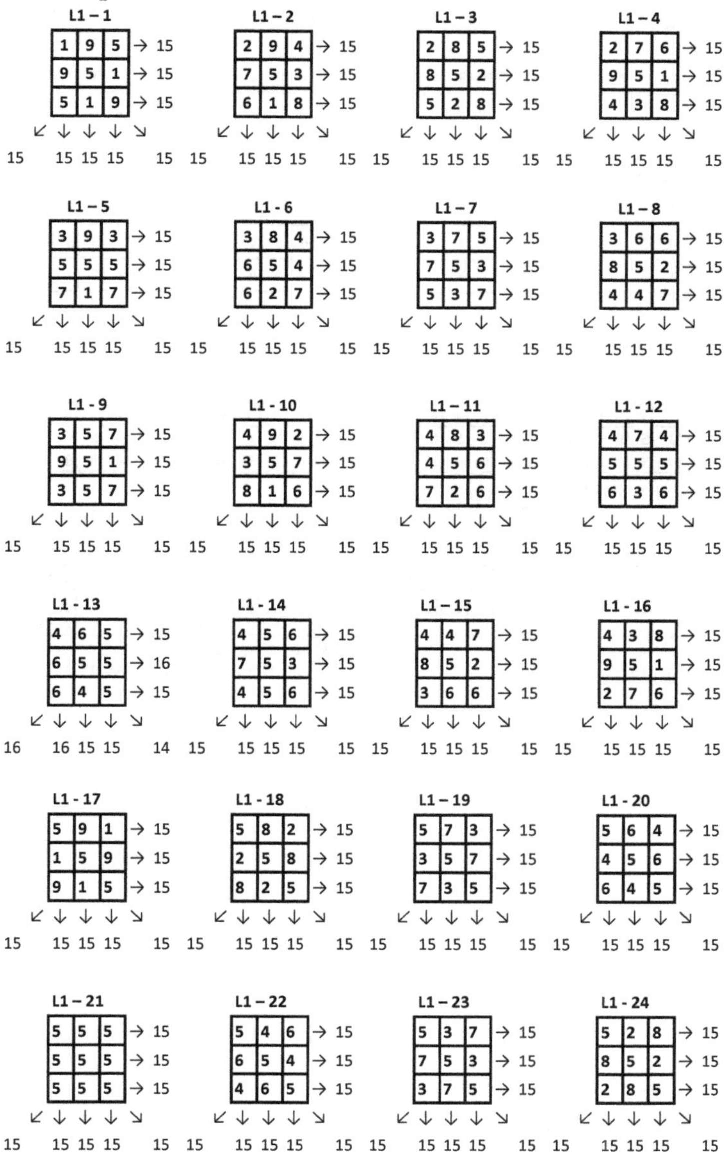

L1 - 25

5	1	9	→ 15
9	5	1	→ 15
1	9	5	→ 15

↙ ↓ ↓ ↓ ↘

15 15 15 15 15

L1 - 26

6	7	2	→ 15
1	5	9	→ 15
8	3	4	→ 15

↙ ↓ ↓ ↓ ↘

15 15 15 15 15

L1 – 27

6	6	3	→ 15
2	5	8	→ 15
7	4	4	→ 15

↙ ↓ ↓ ↓ ↘

15 15 15 15 15

L1 - 28

6	5	4	→ 15
3	5	7	→ 15
6	5	4	→ 15

↙ ↓ ↓ ↓ ↘

15 15 15 15 15

L1 - 29

9	1	5	→ 15
1	5	9	→ 15
5	9	1	→ 15

↙ ↓ ↓ ↓ ↘

15 15 15 15 15

L1 - 30

6	3	6	→ 15
5	5	5	→ 15
4	7	4	→ 15

↙ ↓ ↓ ↓ ↘

15 15 15 15 15

L1 – 31

6	2	7	→ 15
6	5	4	→ 15
3	8	4	→ 15

↙ ↓ ↓ ↓ ↘

15 15 15 15 15

L1 - 32

6	1	8	→ 15
7	5	3	→ 15
2	9	4	→ 15

↙ ↓ ↓ ↓ ↘

15 15 15 15 15

L1 - 33

7	5	3	→ 15
1	5	9	→ 15
7	5	3	→ 15

↙ ↓ ↓ ↓ ↘

15 15 15 15 15

L1 - 34

7	4	4	→ 15
2	5	8	→ 15
6	6	3	→ 15

↙ ↓ ↓ ↓ ↘

15 15 15 15 15

L1 – 35

7	3	5	→ 15
3	5	7	→ 15
5	7	3	→ 15

↙ ↓ ↓ ↓ ↘

15 15 15 15 15

L1 - 36

7	2	6	→ 15
4	5	6	→ 15
4	8	3	→ 15

↙ ↓ ↓ ↓ ↘

15 15 15 15 15

L1 - 37

7	1	7	→ 15
5	5	5	→ 15
3	9	3	→ 15

↙ ↓ ↓ ↓ ↘

15 15 15 15 15

L1 - 38

8	3	4	→ 15
1	5	9	→ 15
6	7	2	→ 15

↙ ↓ ↓ ↓ ↘

15 15 15 15 15

L1 – 39

8	2	5	→ 15
2	5	8	→ 15
5	8	2	→ 15

↙ ↓ ↓ ↓ ↘

15 15 15 15 15

L1 - 40

8	1	6	→ 15
3	5	7	→ 15
4	9	2	→ 15

↙ ↓ ↓ ↓ ↘

15 15 15 15 15

3. All the 40 Possible Order 4 Afamatrix Magic Square Grids.

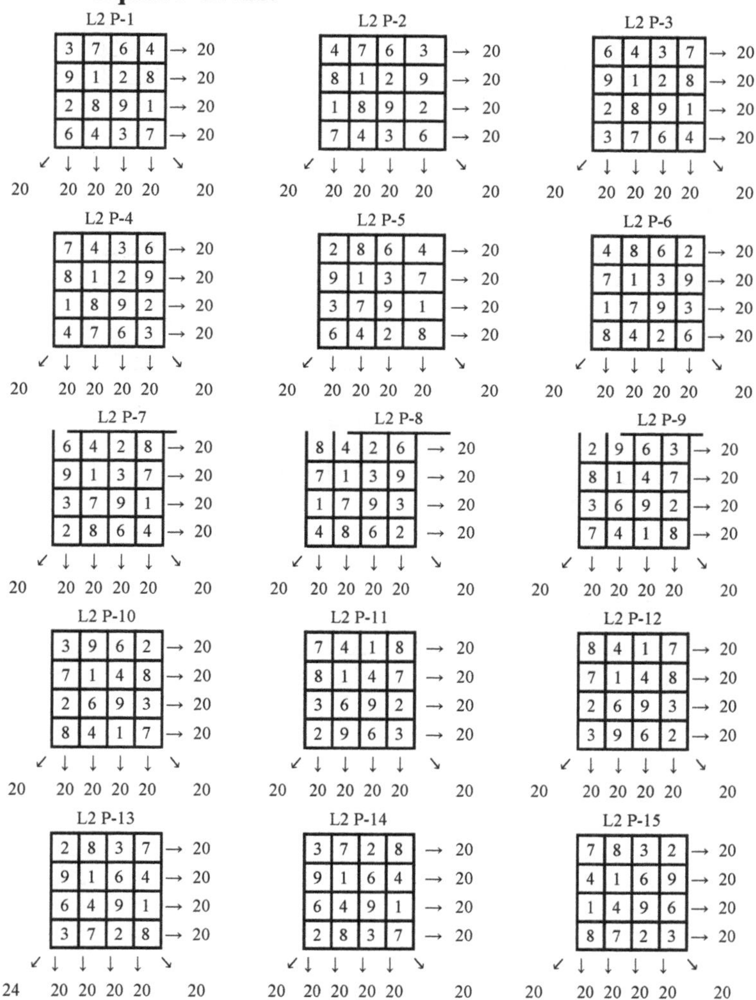

L2 P-1

3	7	6	4	→ 20
9	1	2	8	→ 20
2	8	9	1	→ 20
6	4	3	7	→ 20

↙ 20 ↓ 20 20 20 20 ↘ 20

L2 P-2

4	7	6	3	→ 20
8	1	2	9	→ 20
1	8	9	2	→ 20
7	4	3	6	→ 20

↙ 20 ↓ 20 20 20 20 ↘ 20

L2 P-3

6	4	3	7	→ 20
9	1	2	8	→ 20
2	8	9	1	→ 20
3	7	6	4	→ 20

↙ 20 ↓ 20 20 20 20 ↘ 20

L2 P-4

7	4	3	6	→ 20
8	1	2	9	→ 20
1	8	9	2	→ 20
4	7	6	3	→ 20

↙ 20 ↓ 20 20 20 20 ↘ 20

L2 P-5

2	8	6	4	→ 20
9	1	3	7	→ 20
3	7	9	1	→ 20
6	4	2	8	→ 20

↙ 20 ↓ 20 20 20 20 ↘ 20

L2 P-6

4	8	6	2	→ 20
7	1	3	9	→ 20
1	7	9	3	→ 20
8	4	2	6	→ 20

↙ 20 ↓ 20 20 20 20 ↘ 20

L2 P-7

6	4	2	8	→ 20
9	1	3	7	→ 20
3	7	9	1	→ 20
2	8	6	4	→ 20

↙ 20 ↓ 20 20 20 20 ↘ 20

L2 P-8

8	4	2	6	→ 20
7	1	3	9	→ 20
1	7	9	3	→ 20
4	8	6	2	→ 20

↙ 20 ↓ 20 20 20 20 ↘ 20

L2 P-9

2	9	6	3	→ 20
8	1	4	7	→ 20
3	6	9	2	→ 20
7	4	1	8	→ 20

↙ 20 ↓ 20 20 20 20 ↘ 20

L2 P-10

3	9	6	2	→ 20
7	1	4	8	→ 20
2	6	9	3	→ 20
8	4	1	7	→ 20

↙ 20 ↓ 20 20 20 20 ↘ 20

L2 P-11

7	4	1	8	→ 20
8	1	4	7	→ 20
3	6	9	2	→ 20
2	9	6	3	→ 20

↙ 20 ↓ 20 20 20 20 ↘ 20

L2 P-12

8	4	1	7	→ 20
7	1	4	8	→ 20
2	6	9	3	→ 20
3	9	6	2	→ 20

↙ 20 ↓ 20 20 20 20 ↘ 20

L2 P-13

2	8	3	7	→ 20
9	1	6	4	→ 20
6	4	9	1	→ 20
3	7	2	8	→ 20

↙ 24 ↓ 20 20 20 20 ↘ 20

L2 P-14

3	7	2	8	→ 20
9	1	6	4	→ 20
6	4	9	1	→ 20
2	8	3	7	→ 20

↙ 20 ↓ 20 20 20 20 ↘ 20

L2 P-15

7	8	3	2	→ 20
4	1	6	9	→ 20
1	4	9	6	→ 20
8	7	2	3	→ 20

↙ 20 ↓ 20 20 20 20 ↘ 20

L2 P-16

8	7	2	3	→ 20
4	1	6	9	→ 20
1	4	9	6	→ 20
7	8	3	2	→ 20

↙ ↓ ↓ ↓ ↓ ↘

20 20 20 20 20 20

L2 P-17

2	9	3	6	→ 20
8	1	7	4	→ 20
6	3	9	2	→ 20
4	7	1	8	→ 20

↙ ↓ ↓ ↓ ↓ ↘

20 20 20 20 20 20

L2 P-18

4	7	1	8	→ 20
8	1	7	4	→ 20
6	3	9	2	→ 20
2	9	3	6	→ 20

↙ ↓ ↓ ↓ ↓ ↘

20 20 20 20 20 20

L2 P-19

6	9	3	2	→ 20
4	1	7	8	→ 20
2	3	9	6	→ 20
8	7	1	4	→ 20

↙ ↓ ↓ ↓ ↓ ↘

20 20 20 20 20 20

L2 P-20

8	7	1	4	→ 20
4	1	7	8	→ 20
2	3	9	6	→ 20
6	9	3	2	→ 20

↙ ↓ ↓ ↓ ↓ ↘

20 20 20 20 20 20

L2 P-21

3	9	2	6	→ 20
7	1	8	4	→ 20
6	2	9	3	→ 20
4	8	1	7	→ 20

↙ ↓ ↓ ↓ ↓ ↘

20 20 20 20 20 20

L2 P-22

4	8	1	7	→ 20
7	1	8	4	→ 20
6	2	9	3	→ 20
3	9	2	6	→ 20

↙ ↓ ↓ ↓ ↓ ↘

20 20 20 20 20 20

L2 P-23

6	9	2	3	→ 20
4	1	8	7	→ 20
3	2	9	6	→ 20
7	8	1	4	→ 20

↙ ↓ ↓ ↓ ↓ ↘

20 20 20 20 20 20

L2 P-24

7	8	1	4	→ 20
4	1	8	7	→ 20
3	2	9	6	→ 20
6	9	2	3	→ 20

↙ ↓ ↓ ↓ ↓ ↘

20 20 20 20 20 20

L2 P-25

3	6	7	4	→ 20
9	2	1	8	→ 20
2	9	8	1	→ 20
6	3	4	7	→ 20

↙ ↓ ↓ ↓ ↓ ↘

20 20 20 20 20 20

L2 P-26

4	6	7	3	→ 20
8	2	1	9	→ 20
1	9	8	2	→ 20
7	3	4	6	→ 20

↙ ↓ ↓ ↓ ↓ ↘

20 20 20 20 20 20

L2 P-27

6	3	4	7	→ 20
9	2	1	8	→ 20
2	9	8	1	→ 20
3	6	7	4	→ 20

↙ ↓ ↓ ↓ ↓ ↘

20 20 20 20 20 20

L2 P-28

7	3	4	6	→ 20
8	2	1	9	→ 20
1	9	8	2	→ 20
4	6	7	3	→ 20

↙ ↓ ↓ ↓ ↓ ↘

20 20 20 20 20 20

L2 P-29

4	3	9	4	→ 20
7	2	2	9	→ 20
3	8	8	1	→ 20
6	7	1	6	→ 20

↙ ↓ ↓ ↓ ↓ ↘

20 20 20 20 20 20

L2 P-30

4	3	9	4	→ 20
9	2	2	7	→ 20
1	8	8	3	→ 20
6	7	1	6	→ 20

↙ ↓ ↓ ↓ ↓ ↘

20 20 20 20 20 20

L2 P-31

4	9	3	4	→ 20
7	2	2	9	→ 20
3	8	8	1	→ 20
6	1	7	6	→ 20

L2 P-32

4	9	3	4	→ 20
9	2	2	7	→ 20
1	8	8	3	→ 20
6	1	7	6	→ 20

L2 P-33

6	1	7	6	→ 20
7	2	2	9	→ 20
3	8	8	1	→ 20
4	9	3	4	→ 20

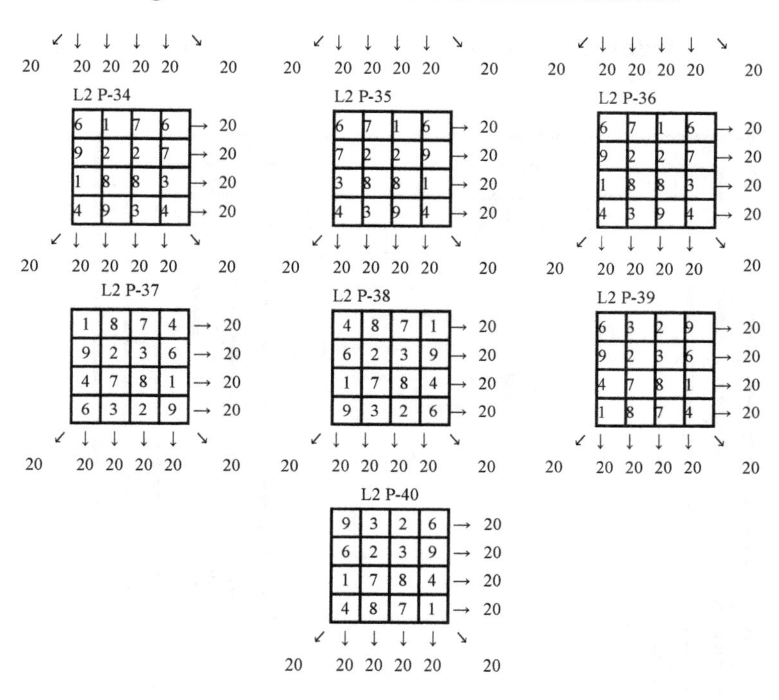

L2 P-34

6	1	7	6	→ 20
9	2	2	7	→ 20
1	8	8	3	→ 20
4	9	3	4	→ 20

↙ ↓ ↓ ↓ ↓ ↘
20 20 20 20 20 20

L2 P-35

6	7	1	6	→ 20
7	2	2	9	→ 20
3	8	8	1	→ 20
4	3	9	4	→ 20

↙ ↓ ↓ ↓ ↓ ↘
20 20 20 20 20 20

L2 P-36

6	7	1	6	→ 20
9	2	2	7	→ 20
1	8	8	3	→ 20
4	3	9	4	→ 20

↙ ↓ ↓ ↓ ↓ ↘
20 20 20 20 20 20

L2 P-37

1	8	7	4	→ 20
9	2	3	6	→ 20
4	7	8	1	→ 20
6	3	2	9	→ 20

↙ ↓ ↓ ↓ ↓ ↘
20 20 20 20 20 20

L2 P-38

4	8	7	1	→ 20
6	2	3	9	→ 20
1	7	8	4	→ 20
9	3	2	6	→ 20

↙ ↓ ↓ ↓ ↓ ↘
20 20 20 20 20 20

L2 P-39

6	3	2	9	→ 20
9	2	3	6	→ 20
4	7	8	1	→ 20
1	8	7	4	→ 20

↙ ↓ ↓ ↓ ↓ ↘
20 20 20 20 20 20

L2 P-40

9	3	2	6	→ 20
6	2	3	9	→ 20
1	7	8	4	→ 20
4	8	7	1	→ 20

↙ ↓ ↓ ↓ ↓ ↘
20 20 20 20 20 20

4. Order 21 Semi-Perfect Afamatrix Magic Square Grid.

																					↗ 105
1	9	1	9	1	9	1	9	1	9	8	4	6	4	6	4	6	4	6	4	3	→ 105
8	2	1	9	1	9	1	9	1	9	9	4	6	4	6	4	6	4	6	4	2	→ 105
2	2	1	9	1	9	1	9	1	9	8	4	6	4	6	4	6	4	3	8	8	→ 105
8	8	8	2	1	9	1	9	1	9	9	4	6	4	6	4	6	4	2	2	2	→ 105
2	2	2	2	1	9	1	9	1	9	8	4	6	4	6	4	3	8	8	8	8	→ 105
8	8	8	8	8	2	1	9	1	9	9	4	6	4	6	4	2	2	2	2	2	→ 105
2	2	2	2	2	2	1	9	1	9	8	4	6	4	3	8	8	8	8	8	8	→ 105
8	8	8	8	8	8	8	2	1	9	9	4	6	4	2	2	2	2	2	2	2	→ 105
2	2	2	2	2	2	2	2	1	9	8	4	3	8	8	8	8	8	8	8	8	→ 105
8	8	8	8	8	8	8	8	8	2	9	4	2	2	2	2	2	2	2	2	2	→ 105
6	7	6	7	6	7	6	7	6	7	5	3	4	3	4	3	4	3	4	3	4	→ 105
3	3	3	3	3	3	3	3	3	6	1	8	7	7	7	7	7	7	7	7	7	→ 105
7	7	7	7	7	7	7	7	7	1	2	6	9	3	3	3	3	3	3	3	3	→ 105
3	3	3	3	3	3	3	6	9	1	1	6	4	8	7	7	7	7	7	7	7	→ 105
7	7	7	7	7	7	7	1	9	1	2	6	4	6	9	3	3	3	3	3	3	→ 105
3	3	3	3	3	6	9	1	9	1	1	6	4	6	4	8	7	7	7	7	7	→ 105
7	7	7	7	7	1	9	1	9	1	2	6	4	6	4	6	9	3	3	3	3	→ 105
3	3	3	6	9	1	9	1	9	1	1	6	4	6	4	6	4	8	7	7	7	→ 105
7	7	7	1	9	1	9	1	9	1	2	6	4	6	4	6	4	6	9	3	3	→ 105
3	6	9	1	9	1	9	1	9	1	1	6	4	6	4	6	4	6	4	8	7	→ 105
7	1	9	1	9	1	9	1	9	1	2	6	4	6	4	6	4	6	4	6	9	→ 105
↓ 105	↓ 105	↓ 105	↓ 105	↓ 105	↓ 105	↓ 105	↓ 105	↓ 105	↓ 105	↓ 105	↓ 105	↓ 105	↓ 105	↓ 105	↓ 105	↓ 105	↓ 105	↓ 105	↓ 105	↓ 105	↘ 105

Note that there are 441 boxes in the grid containing 55 times of each of the digits with the exception of number 5 which is used only once.

Also note that a magic sum of 15, 25, 35, 45, 55, 65, 75, 85, 95 and 105 can be seen in the grid starting from the innermost square.

5. Order 20 Perfected Semi-Imperfect Afamatrix Magic Square Grid.

↗ 100

2	9	1	9	1	9	1	9	6	8	7	3	1	2	8	2	8	2	8	4	→ 100
3	1	9	1	9	1	9	1	9	8	7	2	2	8	2	8	2	8	3	7	→ 100
7	3	2	9	1	9	1	9	6	8	7	3	1	2	8	2	8	4	7	3	→ 100
3	7	3	1	9	1	9	1	9	8	7	2	2	8	2	8	3	7	3	7	→ 100
7	3	7	3	2	9	1	9	6	8	7	3	1	2	8	4	7	3	7	3	→ 100
3	7	3	7	3	1	9	1	9	8	7	2	2	8	3	7	3	7	3	7	→ 100
7	3	7	3	7	3	2	9	6	8	7	3	1	4	7	3	7	3	7	3	→ 100
6	7	6	7	6	7	6	1	9	8	7	2	3	4	3	4	3	4	3	4	→ 100
7	9	7	9	7	9	7	9	2	8	6	4	1	3	1	3	1	3	1	3	→ 100
8	8	8	8	8	8	8	8	9	1	3	7	2	2	2	2	2	2	2	2	→ 100
4	4	4	4	4	4	4	4	3	7	9	1	6	6	6	6	6	6	6	6	→ 100
3	1	3	1	3	1	3	1	6	4	2	8	9	7	9	7	9	7	9	7	→ 100
4	4	4	4	4	4	4	7	1	2	3	8	9	6	6	6	6	6	6	6	→ 100
4	6	4	6	4	6	6	1	4	2	3	7	9	8	4	6	4	6	4	6	→ 100
6	4	6	4	6	7	1	9	1	2	3	8	8	2	9	4	6	4	6	4	→ 100
4	6	4	6	6	1	9	1	4	2	3	7	9	8	2	8	4	6	4	6	→ 100
6	4	6	7	1	9	1	9	1	2	3	8	8	2	8	2	9	4	6	4	→ 100
4	6	6	1	9	1	9	1	4	2	3	7	9	8	2	8	2	8	4	6	→ 100
6	7	1	9	1	9	1	9	1	2	3	8	8	2	8	2	8	2	9	4	→ 100
6	1	9	1	9	1	9	1	4	2	3	7	9	8	2	8	2	8	2	8	→ 100
↓ 100	↓ 100	↓ 100	↓ 100	↓ 100	↓ 100	↓ 100	↓ 100	↓ 100	↓ 100	↓ 100	↓ 100	↓ 100	↓ 100	↓ 100	↓ 100	↓ 100	↓ 100	↓ 100	↓ 100	↘ 100

Note that here are 400 boxes in the grid containing 50 times of each of the digits with the exception of number 5 which is not used at all. Also note that a magic sum of 20, 30, 40, 50, 60, 70, 80, 90 and 100 can be seen in the grid starting from the innermost square. Also note that the imperfect Magic Square Grids is perfected by the preceding Magic Square Grid inside the Main Magic Square Grid.

6. Order 18 Multi-Sectional Odd Afamatrix Magic Square Grid.

																		↗ 90
2	9	4	2	7	6	4	9	2	6	7	2	6	1	8	8	3	4	→ 90
7	5	3	9	5	1	3	5	7	1	5	9	7	5	3	1	5	9	→ 90
6	1	8	4	3	8	8	1	6	8	3	4	2	9	4	6	7	2	→ 90
2	9	4	2	7	6	4	9	2	6	7	2	6	1	8	8	3	4	→ 90
7	5	3	9	5	1	3	5	7	1	5	9	7	5	3	1	5	9	→ 90
6	1	8	4	3	8	8	1	6	8	3	4	2	9	4	6	7	2	→ 90
2	9	4	2	7	6	4	9	2	6	7	2	6	1	8	8	3	4	→ 90
7	5	3	9	5	1	3	5	7	1	5	9	7	5	3	1	5	9	→ 90
6	1	8	4	3	8	8	1	6	8	3	4	2	9	4	6	7	2	→ 90
2	9	4	2	7	6	4	9	2	6	7	2	6	1	8	8	3	4	→ 90
7	5	3	9	5	1	3	5	7	1	5	9	7	5	3	1	5	9	→ 90
6	1	8	4	3	8	8	1	6	8	3	4	2	9	4	6	7	2	→ 90
2	9	4	2	7	6	4	9	2	6	7	2	6	1	8	8	3	4	→ 90
7	5	3	9	5	1	3	5	7	1	5	9	7	5	3	1	5	9	→ 90
6	1	8	4	3	8	8	1	6	8	3	4	2	9	4	6	7	2	→ 90
2	9	4	2	7	6	4	9	2	6	7	2	6	1	8	8	3	4	→ 90
7	5	3	9	5	1	3	5	7	1	5	9	7	5	3	1	5	9	→ 90
6	1	8	4	3	8	8	1	6	8	3	4	2	9	4	6	7	2	→ 90
↓ 90	↓ 90	↓ 90	↓ 90	↓ 90	↓ 90	↓ 90	↓ 90	↓ 90	↓ 90	↓ 90	↓ 90	↓ 90	↓ 90	↓ 90	↓ 90	↓ 90	↓ 90	↘ 90

Note that the Magic Sum of 90 in the entire grid, Magic Sum of 60 in the Bigger Inner Section and Magic Sum of 30 in the 13 Smaller sections while the entire grid is made up of 36 magic squares with Magic sum of 15.

7. **Order 24 Multi-Sectional Even Afamatrix Magic Square Grid.**

↗ 120

3	7	6	4	4	7	6	3	6	4	3	7	4	8	1	7	6	9	2	3	7	8	1	4	→ 120
9	1	2	8	8	1	2	9	9	1	2	8	7	1	8	4	4	1	8	7	4	1	8	7	→ 120
2	8	9	1	1	8	9	2	2	8	9	1	6	2	9	3	3	2	9	6	3	2	9	6	→ 120
6	4	3	7	7	4	3	6	3	7	6	4	3	9	2	6	7	8	1	4	6	9	2	3	→ 120
7	4	3	6	2	8	6	4	4	8	6	2	3	6	7	4	4	6	7	3	6	3	4	7	→ 120
8	1	2	9	9	1	3	7	7	1	3	9	9	2	1	8	8	2	1	9	9	2	1	8	→ 120
1	8	9	2	3	7	9	1	1	7	9	3	2	9	8	1	1	9	8	2	2	9	8	1	→ 120
4	7	6	3	6	4	2	8	8	4	2	6	6	3	4	7	7	3	4	6	3	6	7	4	→ 120
6	4	2	8	8	4	2	6	2	9	6	3	7	3	4	6	4	3	9	4	4	3	9	4	→ 120
9	1	3	7	7	1	3	9	8	1	4	7	8	2	1	9	7	2	2	9	9	2	2	7	→ 120
3	7	9	1	1	7	9	3	3	6	9	2	1	9	8	2	3	8	8	1	1	8	8	3	→ 120
2	8	6	4	4	8	6	2	7	4	1	8	4	6	7	3	6	7	1	6	6	7	1	6	→ 120
3	9	6	2	7	4	1	8	8	4	1	7	4	9	3	4	4	9	3	4	6	1	7	6	→ 120
7	1	4	8	8	1	4	7	7	1	4	8	7	2	2	9	9	2	2	7	7	2	2	9	→ 120
2	6	9	3	3	6	9	2	2	6	9	3	3	8	8	1	1	8	8	3	3	8	8	1	→ 120
8	4	1	7	2	9	6	3	3	9	6	2	6	1	7	6	6	1	7	6	4	9	3	4	→ 120
2	8	3	7	3	7	2	8	7	8	3	2	6	1	7	6	6	7	1	6	6	7	1	6	→ 120
9	1	6	4	9	1	6	4	4	1	6	9	9	2	2	7	7	2	2	9	9	2	2	7	→ 120
6	4	9	1	6	4	9	1	1	4	9	6	1	8	8	3	3	8	8	1	1	8	8	3	→ 120
3	7	2	8	2	8	3	7	8	7	2	3	4	9	3	4	4	3	9	4	4	3	9	4	→ 120
8	7	2	3	2	9	3	6	4	7	1	8	1	8	7	4	4	8	7	1	6	3	2	9	→ 120
4	1	6	9	8	1	7	4	8	1	7	4	9	2	3	6	6	2	3	9	9	2	3	6	→ 120
1	4	9	6	6	3	9	2	6	3	9	2	4	7	8	1	1	7	8	4	4	7	8	1	→ 120
7	8	3	2	4	7	1	8	2	9	3	6	6	3	2	9	9	3	2	6	1	8	7	4	→ 120
↓ 120	↓ 120	↓ 120	↓ 120	↓ 120	↓ 120	↓ 120	↓ 120	↓ 120	↓ 120	↓ 120	↓ 120	↓ 120	↓ 120	↓ 120	↓ 120	↓ 120	↓ 120	↓ 120	↓ 120	↓ 120	↓ 120	↓ 120	↓ 120	↘ 120

Note that the Magic Sum of 120 in the entire grid, Magic Sum of 80 in the Bigger Inner Section, Magic Sum of 60 in the 4 Quadrants of the Big Magic Square Grid and Magic Sum of 20 in the 9 Smaller sections while the entire grid is made up of 36 magic squares with Magic sum of 20.

CHAPTER 6

The Hyper-Afamatrix Magic Square Grid Adaptation

The definitions and explanations so far refer to the first category of Afamatrix Magic Square Grids that can best be described as Simple or Normal Afamatrix Magic Square Grids. There is another important Major category. This category is known as Hyper-Afamatrix Magic Square Grids. This category is not limited to the use of single digits only in the formation. Hyper-Afamatrix Magic Square Grid is further divided into two Categories namely; Classical-Afamatrix Magic Square Grids and Hybrid Afamatrix Magic Square Grids.

Classical Hyper-Afamatrix Magic Square Grids.

In the ***Classical version*** of Hyper-Afamatrix Magic Square Grids, either single or more digits are used in the initial formation.

Hybrid Hyper-Afamatrix Magic Square Grids.

In the ***Hybrid version*** of Hyper-Afamatrix Magic Square Grids the numbers are derived or adapted from a Simple Afamatrix Magic Square Grid when Arithmetic operations like Addition, Subtraction, Multiplication, Division, etc are applied.
The second Magic Square Grid below shows an Example of a Hybrid Hyper-Afamatrix Magic Square Grid which is formed by the addition of 2 to every digit in the first Magic Square Grid.

	1	6	8	7	3	→ 25
	2	2	9	4	8	→ 25
	6	7	5	3	4	→ 25
	9	6	1	8	1	→ 25
	7	4	2	3	9	→ 25
/	↓	↓	↓	↓	↓	\
25	25	25	25	25	25	25

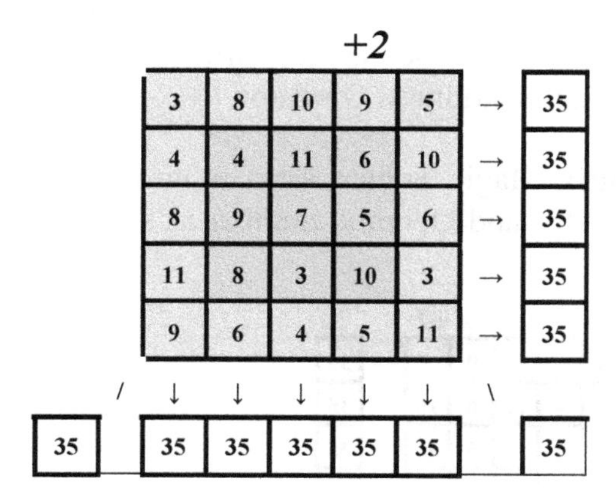

+2

	3	8	10	9	5	→ 35
	4	4	11	6	10	→ 35
	8	9	7	5	6	→ 35
	11	8	3	10	3	→ 35
	9	6	4	5	11	→ 35
/	↓	↓	↓	↓	↓	\
35	35	35	35	35	35	35

Operands & Operator of a Hybrid Hyper-Afamatrix Magic Square Grid.

In Arithmetic, Operators are symbols which uses one or more Operands or expressions to perform arithmetic Operations (e.g. +, - , / ,etc) while Operands are variables or expressions, which are used with Operators to form an expression. Combination of operands and operators form an expression.

In Afamatrix, Combination of the Operand Magic Square Grid, the Operand Number and an Operator forms the Hybrid Hyper-Afamatrix Magic Square Grid.

The first Magic Square Grid above is the Operand Magic Square Grid which is used in combination with the Operand Number *2* to form the second Magic Square Grid after an Arithmetic Operation using the Addition Operator.

A Hybrid Hyper-Afamatrix Magic Square Grid is normally represented with the Operator and Operand number as shown below;

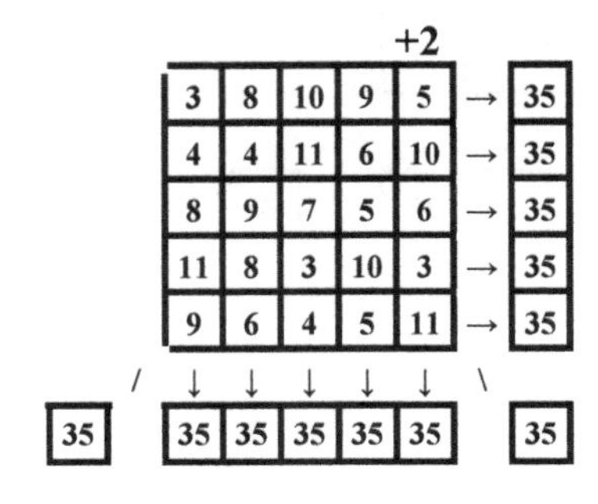

Examples of Hybrid Hyper-Afamatrix Magic Grids formed from a Simple Afamatrix Magic Square Grid by the Application of different Arithmetic Operations.

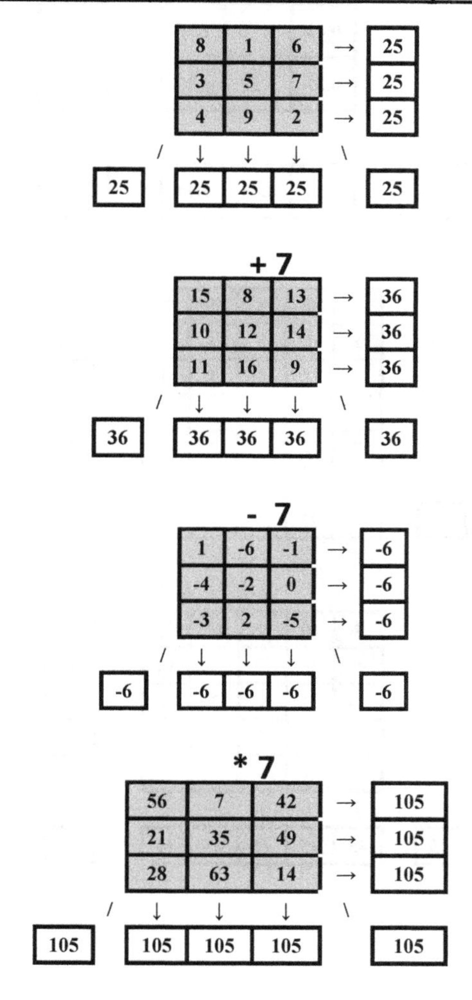

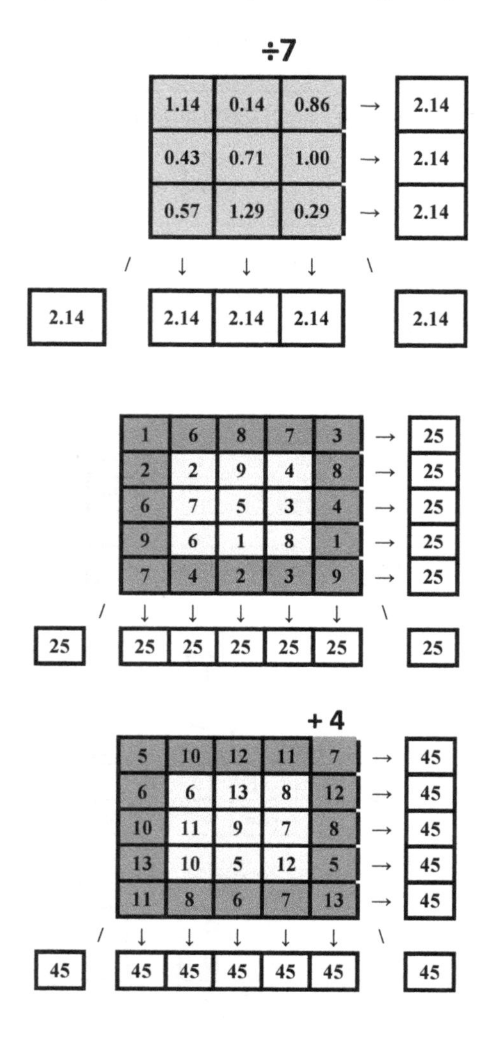
÷7
1.14
0.14
0.86
2.14
0.43
0.71
1.00
2.14
0.57
1.29
0.29
2.14
2.14
2.14
2.14
2.14
2.14
1
6
8
7
3
25
2
2
9
4
8
25
6
7
5
3
4
25
9
6
1
8
1
25
7
4
2
3
9
25
25
25
25
25
25
25
25
+4
5
10
12
11
7
45
6
6
13
8
12
45
10
11
9
7
8
45
13
10
5
12
5
45
11
8
6
7
13
45
45
45
45
45
45
45
45

-4

-3	2	4	3	-1	→	5
-2	-2	5	0	4	→	5
2	3	1	-1	0	→	5
5	2	-3	4	-3	→	5
3	0	-2	-1	5	→	5

/	↓	↓	↓	↓	↓	\
5	5	5	5	5	5	5

***4**

4	24	32	28	12	→	100
8	8	36	16	32	→	100
24	28	20	12	16	→	100
36	24	4	32	4	→	100
28	16	8	12	36	→	100

/	↓	↓	↓	↓	↓	\
100	100	100	100	100	100	100

÷4

0.25	1.50	2.00	1.75	0.75	→	6.25
0.50	0.50	2.25	1.00	2.00	→	6.25
1.50	1.75	1.25	0.75	1.00	→	6.25
2.25	1.50	0.25	2.00	0.25	→	6.25
1.75	1.00	0.50	0.75	2.25	→	6.25

/	↓	↓	↓	↓	↓	\
6.25	6.25	6.25	6.25	6.25	6.25	6.25

Perfect and Imperfect Hyper-Afamatrix Magic Square Grids

All the Hyper-Afamatrix Magic Square Grids that obey the complete rule of Afamatrix before a mathematical operation is applied for it's formation are classified as Perfect Hyper-Afamatrix Magic Square Grids.

	11	19	18	14	13	→	75
	13	12	19	14	17	→	75
	16	17	15	13	14	→	75
	18	16	11	18	12	→	75
	17	11	12	16	19	→	75
↙	↓	↓	↓	↓	↓	↘	
75	75	75	75	75	75		75

Example of a Semi-Perfect Hyper-Afamatrix Magic Square Grid
(Note that the numbers are used equally except number 15 that is used only once.)

Imperfect Hyper-Afamatrix Magic Square Grids comprises of all the Afamatrix Magic Square Grids that do not fully obey the complete rule of Afamatrix before a mathematical operation is applied for it's formation.

These include the Hyper-Afamatrix Magic Square Grids that do not contain an equal amount of each integer used in the entire grid. Though Imperfect Hyper-Afamatrix Magic Square Grids also forms an integral and interesting part of Afamatrix,

I will pay more attention to Perfect Hyper-Afamatrix Magic Square Grids for our study, for the sake of convenience.

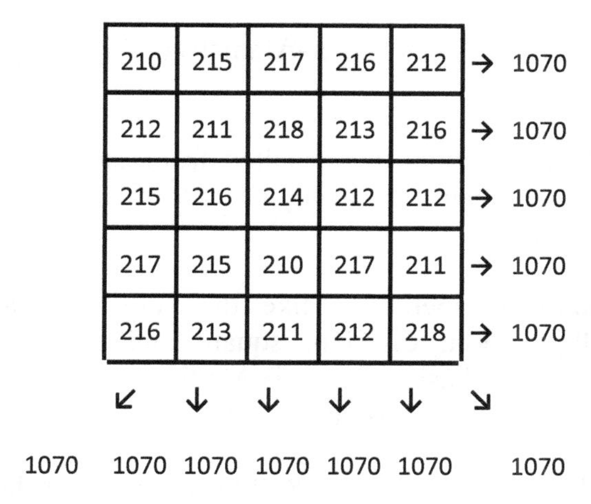

210	215	217	216	212	→ 1070
212	211	218	213	216	→ 1070
215	216	214	212	212	→ 1070
217	215	210	217	211	→ 1070
216	213	211	212	218	→ 1070
↙ 1070 (diagonal) ; ↓ 1070	↓ 1070	↓ 1070	↓ 1070	↓ 1070	↘ 1070

Example of an Imperfect Hyper-Afamatrix Magic Square Grid.
(Note that the numbers are not used equally; number 210 is used two times while number 212 is used four times, the rest of the numbers are used three times each)

Rules of Perfect Hyper-Afamatrix Magic Square Grids

The rules of a Perfect Hyper-Afamatrix Magic Square Grid include:

1) Only the figures that resulted from the application of arithmetic operation to digit 1 to 9 of a basic Afamatrix square are used.

   Arithmetic operations like the addition, subtraction, multiplication, division, square and/or cube can be applied to a basic Afamatrix Magic Square Grid in order to obtain a hyper-Afamatrix Magic Square Grid.

2) The digits 1, 2, 3, 4, 6, 7, 8, 9 or the result of the applied arithmetic operation, are used equally except digit 5 or the result of its arithmetic operation with any chosen number which is not used or used only once in every Hyper-Afamatrix square.

CHAPTER 7

Part 1 - Afam's Method of Constructing Magic Square Grids

Number Ranges and Components of Magic Square Grids.

I wish to introduce to you a new method of constructing Magic Square Grids. I called it "The Afam's Method", named after my Igbo name. This method can be used in constructing any type of Magic Square Grid. It can also be used in generating all the possible Variations of any Magic Square Grid.

The Afam's method was formulated as a result of my quest to find a centralized or universal representation of Magic Square Grids, and a Construction Technique that will be Common to all Magic Square Grids. Two major aspects where then considered; One is to disintegrate the Magic Square Grid into generalized entities or Component Parts and to discover the relationship between these component parts, the second is to find out about the underlying Mathematical Principles guiding the Magic Square Grids.

Afam's method involves the use of **Magic Square Grid Tables** and **Checking Tables** in the Construction of Magic Square Grids. The method can be very simple as well as complex depending on the type of Magic Square Grid being formed. Basically, the method involves spreading out the cells of the Magic Square Grid in a single row such that every cell in the row represents a cell in the Magic Square Grid, the row is then used to form a table of possible variations of numbers known as the **Magic Square Grid Table.** Another Table, known as the **Checking Table**is inserted by the side of the

Magic Square Table and used in checking the numbers on each of the rows on the expected requirement of the Magic Square Grid, like Magic Sum, number of digits, etc. The Rows whose different summations form a Magic Square Grid are selected while those that do not comply are filtered out and deleted.

In the case of Order 3 Magic Square Grid, the row will contain 9 cells representing the 9 cells of the Grid. The number of rows generated will usually get too large if any constraint is not applied. The essence of Afam's method is to show how constraints can be applied in order to filter out the unwanted rows, thereby reducing the number of rows generated. Without applying any constraints, the numbers starting from 000000000 to 999999999 per row will need to be **incrementally** filled-in. This is illustrated with the order 3 Magic Square Grid Table below.

Other types of **Checking Tables**can also be created in order to check the other Properties of the Magic Square Grids. These include; the number count of each digit used in the Magic Square Grid, the Range of Possible Magic Square Sums etc.

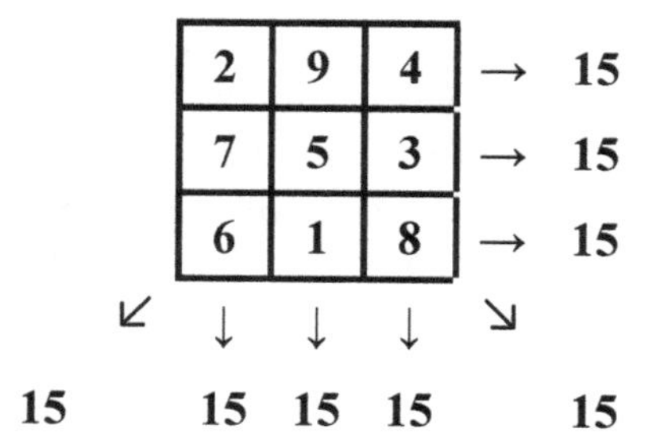

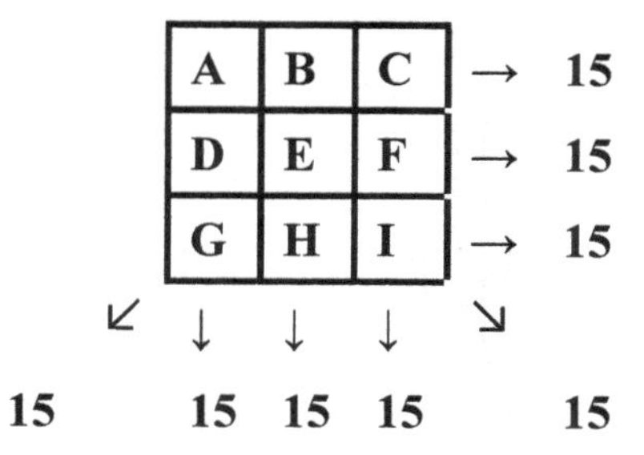

MAGIC SQUARE TABLE									SUMMATION CHECKING TABLE								
									ROWS			COLUMN S			DIAGONA LS		CHEC K
A	B	C	D	E	F	G	H	I	A + B + C	D + E + F	G + H + I	A + D + G	B + E + H	C + F + I	A + E + I	C + E + G	OK or NOK
0	0	0	0	0	0	0	0	0	0	0	0	0	0	0	0	0	NOK
0	0	0	0	0	0	0	0	1	0	0	1	0	0	1	1	0	NOK
0	0	0	0	0	0	0	0	2	0	0	2	0	0	2	2	0	NOK
0	0	0	0	0	0	0	0	3	0	0	3	0	0	3	3	0	NOK
0	0	0	0	0	0	0	0	4	0	0	4	0	0	4	4	0	NOK
0	0	0	0	0	0	0	0	5	0	0	5	0	0	5	5	0	NOK
0	0	0	0	0	0	0	0	6	0	0	6	0	0	6	6	0	NOK
0	0	0	0	0	0	0	0	7	0	0	7	0	0	7	7	0	NOK
9	9	9	9	9	9	9	9	2	27	27	20	27	27	20	20	27	NOK

9	9	9	9	9	9	9	9	3
9	9	9	9	9	9	9	9	4
9	9	9	9	9	9	9	9	5
9	9	9	9	9	9	9	9	6
9	9	9	9	9	9	9	9	7
9	9	9	9	9	9	9	9	8
9	9	9	9	9	9	9	9	9

27	**27**	**21**	**27**	**27**	**21**	**21**	**27**	**NOK**
27	**27**	**22**	**27**	**27**	**22**	**22**	**27**	**NOK**
27	**27**	**23**	**27**	**27**	**23**	**23**	**27**	**NOK**
27	**27**	**24**	**27**	**27**	**24**	**24**	**27**	**NOK**
27	**27**	**25**	**27**	**27**	**25**	**25**	**27**	**NOK**
27	**27**	**26**	**27**	**27**	**26**	**26**	**27**	**NOK**
27	**27**	**27**	**27**	**27**	**27**	**27**	**27**	**NOK**

Ordinarily, using the Method as described above could have been good enough if creating a table with extremely large number of rows is easy, but practically, it is almost impossible because of the number of rows involved. In the case of Order 3 Magic Square Grid as shown above, Nine Hundred and Ninety Nine Million, Nine Hundred and Ninety Nine Thousand, Nine Hundred and Ninety Nine Rows of numbers will need to be generated. (999,999,999).

For the method to be used as described above on all Magic Square Grid Orders, a table containing billions or trillions of rows of numbers will need to be created before the compliant rows can be selected.

Therefore, we need to apply some constraints in order to narrow down the size of table to be created. These constraints help in filtering out the unwanted rows. The constraints are applied by first of all specifying the Number Range to be used, and by disintegrating the Magic Square Grid into Component Parts in other to fill in the numbers that best fits in. The essence of disintegrating the Magic Square Grids into Components parts and the study of its underlying

Mathematical principles is to enable us have a proper understanding of the relationship that exist between numbers contained in the different variations of Magic square Grids, as such the unwanted numbers can be filtered out.

The Number Range of a Magic Square Grid

One way of applying constraints in order to reduce the number of rows involved while generating Magic Square Grids Tables is to specify the Number Range to be used in constructing the Magic Square Grid. Specifying the Number Range is one way that helps so much in narrowing down the enormous data that would have been generated.

Types of Number Range Specifications

Magic Square Grid Number Range specification can be of two types namely **Inclusive** or **Exclusive**.

In inclusive Number Range application, only the Minimum Number limit and Maximum number limit are specified. While in exclusive Number Range, each of the numbers to be used in the Magic Square Grid are specified. If inclusive number range is specified, then any of the numbers that fall within the specified range can be used in constructing the Magic Square Grid, but if Exclusive Number range is specified, then only the chosen numbers can be used in constructing the Magic Square Grid.

For instance, if an Inclusive Number Range of 1 to 50 is specified, then all the numbers starting from number 1 to number 50 can be used in constructing the Magic Square Grid. On the other hand, if an Exclusive Number Range of 1-50 is specified, then specific numbers within the range of 1-50 has to be mentioned. For example, the even numbers within 1 to 50 can be exclusively specified. Likewise, numbers 1, 4, 6, 9, 17, 18, 35, 36, 39, 47, 50 only can be selected exclusively as the specification.

Inclusive Number Ranges are best generated **Incrementally** while Exclusive number Ranges are best generated by **Permutation**. The Magic Square Grid Table in the section above shows an example of an Inclusive Number Range Specification where the numbers are generated Incrementally. The Magic Square Grid Table in the section below shows an example Exclusive Number Range specification where the number range is generated by Permutation.

Number Repeat and Exceptions in a Range

Before now, every digit or number is used once in a normal Magic Square Grid, but with my new discoveries in Magic Square Grids, The number of times each of the digits that make up a Perfect Magic Square Grid is repeatedly used in constructing a Magic square Grid has to be specified.In Perfect Magic Square Grids, the digits or numbers are used equally, although there are exceptions which is also specified. The exceptions could be that a particular digit or number should be used specific number of times which is different from the rest of the numbers.

The Basic Magic Square Grid Parameters Table

The constraints applied on any Magic Square Grid are shown in the Magic Square Grid Parameters Table. The Table contains all the information with which a Magic Square Grid can be constructed.

The **Basic Magic Square Grid Parameters** Table is a minimized version of the Magic Square Grid Parameters Table, it contains the Minimal Parameters that can be specified for a Magic Square Grid. The minimal specifications includes; The Magic Square Grid Order, The expected Magic Sum and the Range of Numbers to be used. As we proceed, more parameters will be added in order to introduce more Constraints or restrictions. This will in turn reduce the number of Permutations, Compositions and/or group of numbers that needs to be generated. The examples shown below is for an Order 3 Magic Square Grid.

BASIC MAGIC SQUARE GRID PARAMETERS TABLE		
S/N	**PARAMETERS**	**VALUE**
1	**Order / Level**	3 / 1
2	**Magic Sum**	15
3	**Type of Number Range**	Exclusive
4	**Number Range**	1,2,3,4,5,6,7,8, 9
6	**RepeatableNumbers**	No
7	**Maximum Repeat Quantity**	1
8	**Number Range Exceptions**	Nil
9	**Exceptions Quantity**	Nil

Given the Parameters in the Basic Magic Square Grid Parameters Table above, the first number in the Magic Square Grid Table will be 1,2,3,4,5,6,7,8,9, this is because the number range has been specified to be exclusively 1,2,3,4,5,6,7,8,9. Permutations of the first number are then generated in order to fill in the remaining rows.

Understanding Permutations

" In mathematics, it is known that the number of permutations of a string is equal to the factorial of the length of the string. For example, the word RED has a length of three -- which means that the letters can be rearranged in six different ways: RED, RDE, ERD, EDR, DRE, and DER.

Therefore, The number of permutations of *n* distinct objects is $n\times (n - 1) \times (n - 2)\times\cdots\times 2\times 1$, This number is called "*n* factorial" and written as "*n*!".

In a case where the number of chosen strings for Permutation is less than the available strings, the Permutation formula comes into play.
i.e. Permutation Pn, k = n! / (n-k)! Where;
n=Number of Elements &
k= number of Chosen_Elements
and if n= k
then, Permutation Pn,k = Factorial n!

For instance, as shown in the Basic Magic Square Grid Parameters Table above;
Number of Elements = 9
Number of Chosen_Elements =9

Permutation P9,9 = Factorial 9! =362,880
Therefore; 362,880 Permutations is expected to be generated.

MAGIC SQUARE TABLE										**SUMMATION CHECKING TABLE**								
										ROWS			**COLUMNS**			**DIAGONALS**		**CHECK**
S/N	**A**	**B**	**C**	**D**	**E**	**F**	**G**	**H**	**I**	**A+B+C**	**D+E+F**	**G+H+I**	**A+D+G**	**B+E+H**	**C+F+I**	**A+E+I**	**C+E+G**	OK Or NOK
1	1	2	3	4	5	6	7	8	9	**6**	**15**	**24**	**12**	**15**	**18**	**15**	**15**	**NOK**
2	1	2	3	4	5	6	7	9	8	**6**	**15**	**24**	**12**	**16**	**17**	**14**	**15**	**NOK**
3	1	2	3	4	5	6	8	7	9	**6**	**15**	**24**	**13**	**14**	**18**	**15**	**16**	**NOK**
4	1	2	3	4	5	6	8	9	7	**6**	**15**	**24**	**13**	**16**	**16**	**13**	**16**	**NOK**
5	1	2	3	4	5	6	9	7	8	**6**	**15**	**24**	**14**	**14**	**17**	**14**	**17**	**NOK**
6	1	2	3	4	5	6	9	8	7	**6**	**15**	**24**	**14**	**15**	**16**	**13**	**17**	**NOK**
7	1	2	3	4	5	7	6	8	9	**6**	**16**	**23**	**11**	**15**	**19**	**15**	**14**	**NOK**
8	1	2	3	4	5	7	6	9	8	**6**	**16**	**23**	**11**	**16**	**18**	**14**	**14**	**NOK**
362873	9	8	7	6	5	3	4	1	2	**24**	**14**	**7**	**19**	**14**	**12**	**16**	**16**	**NOK**
362874	9	8	7	6	5	3	4	2	1	**24**	**14**	**7**	**19**	**15**	**11**	**15**	**16**	**NOK**
362875	9	8	7	6	5	4	1	2	3	**24**	**15**	**6**	**16**	**15**	**14**	**17**	**13**	**NOK**
362876	9	8	7	6	5	4	1	3	2	**24**	**15**	**6**	**16**	**16**	**13**	**16**	**13**	**NOK**
362877	9	8	7	6	5	4	2	1	3	**24**	**15**	**6**	**17**	**14**	**14**	**17**	**14**	**NOK**
362878	9	8	7	6	5	4	2	3	1	**24**	**15**	**6**	**17**	**16**	**12**	**15**	**14**	**NOK**
362879	9	8	7	6	5	4	3	1	2	**24**	**15**	**6**	**18**	**14**	**13**	**16**	**15**	**NOK**
362880	9	8	7	6	5	4	3	2	1	**24**	**15**	**6**	**18**	**15**	**12**	**15**	**15**	**NOK**

As you have seen, just specifying the number range for the order 3 Magic Square Grid has brought down the number of rows in the table from **999,999,999** to **362,880** out of which only **8** Perfect Magic Square Grids can be formed.

If these numerous number of Permutations can be generated in just "Order 3" which has only 9 Cells, You can imagine how many permutations the Order 4 that has 16 Cells can generate **20,922,789,888,000**. Not even considering Order 5 **15,511,210,043,330,985,984,000,000** and other Higher Magic Square Grid Orders. Therefore, we need to introduce more constraints for us to drastically reduce the number of permutations generated.

For us to introduce more constraints, we need to specify more parameters other than the basic parameters shown in the Basic Magic Square Grid Parameters Table, and to be able to specify these parameters, we have to fragment the Magic Square Grids into component parts as mentioned in the beginning. This will lead us into making a more **Comprehensive Magic Square Grid Parameters Table** containing all the Parameters that have to be specified for a Magic Square Grid.

In the ensuing section, we will find out about the Components Parts in which I have disintegrated the Magic Square Grids, we will also find out about the underlying mathematical principles concerning these component parts and how these Mathematical principles can be applied in constructing Magic Square Grids.

Components of Magic Square Grids

As you, probably have known, different types of Magic Square Grids are unique because of their unique type of number arrangements. In this section, we will look at the different components of Magic Square Grids that makes the unique number arrangements possible. I disintegrated the Magic Square Grid into different Components such that it can be universally applied. Magic Square Grids are made up of Four (4) Minor and Two (2) Major distinct Components. The constructions of the different unique variations of Magic Square Grids are dependent on the proper application and manipulation of these distinct components. This makes it possible to create endless variations of Magic Square Grids. A unique set of number arrangements in a particular Grid can be manipulated to form a different unique Magic Square Grid or Hyper- Magic Square Grid if the distinct components are properly understood.

We will first of all have an in-depth look at the different components of Magic Square Grids, after which we will look at the Interrelation of these Components in the formation of different variations of Magic square Grids.

Minor Components of Magic Square Grids are;

There are Four (4) Minor Component in every Magic Square Grid, they are;

1. The Cells or Boxes
2. The Rows
3. The Columns
4. The Diagonals
    a. Forward Slash Diagonal
    b. Backward Slash Diagonal

Major Components of Magic Square Grids are;

There are Two Major Component in every Magic Square Grid, they are; the Frame Component and The Sectional Component.

The Major Component is Sub-divided as follows;

1) **Frame Component**
   a) "X" Component
   b) "+Plus" Component
      i) "Vertical +Plus" Component (Vn-2)
         (1) Top Side "Vertical +Plus" Component (TVn-2)
         (2) Down Side "Vertical +Plus" Component (DVn-2)

      ii) "Horizontal +Plus" Component (Hn-2)
         (1) Left Side Horizontal Plus Component (LHn-2)
         (2) Right Side Horizontal Plus Component (RHn-2)

2) **The Sectional Components**
   a) Quad
   b) Ninth
   c) Triangular
   Etc

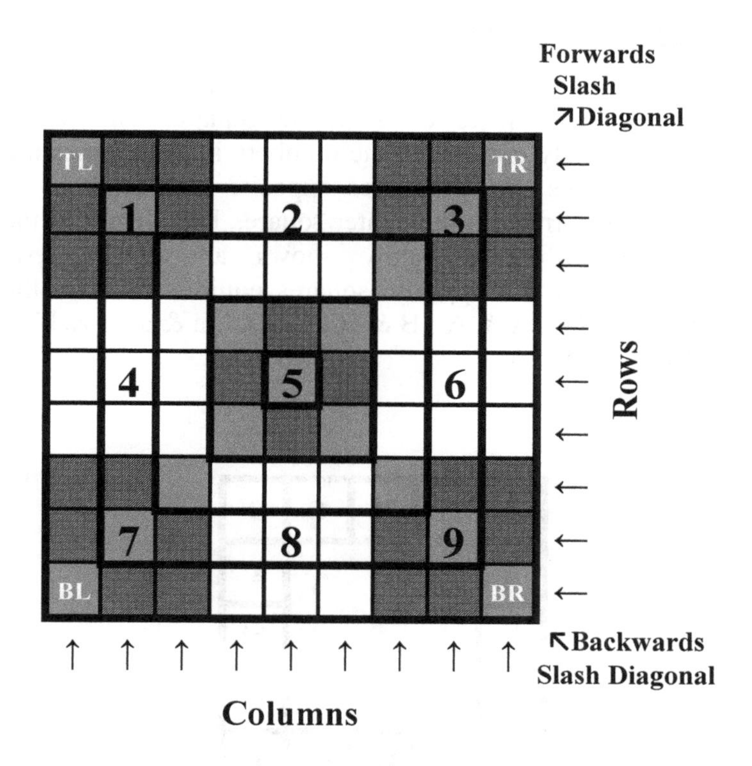

Diagram showing the Rows, Columns, Diagonals, Frames and Sectional Components of a Magic Square Grid

The Cells, Rows, Columns and Diagonal Components

The Cells, Rows, Columns and Diagonal Components make up the Minor Component of a Magic Square Grid. The Diagonal Component is made up of the Forward Slash Diagonal The and Backward Slash Diagonal.

The Frame Component

This is the layered set of squares in an equidistant pattern from the center square of a Magic Square. The layers continue endlessly outwards. The numbers in the Frames are checked in terms of the Squares opposite each other in an equidistant pattern from the center square. The arrows in the Magic Square Frame below shows an alphanumeric representation of the opposite squares equidistant from the center square. I.e. (A & A', B & B', C & C', a & a', b & b', c & c', 1 & 1', 2 & 2')

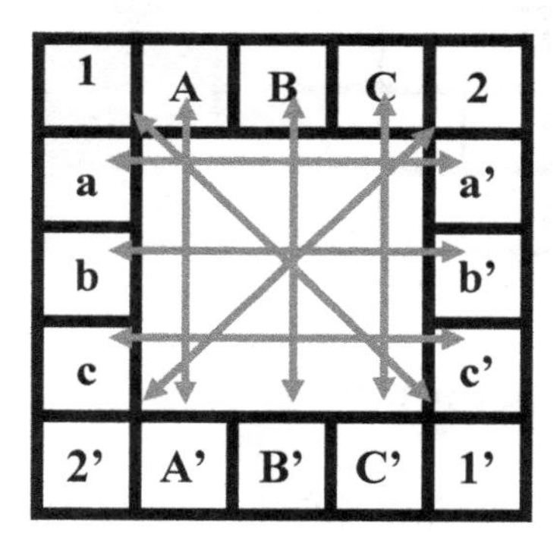

Use of Numbers in the Frames

The numbers in the Squares opposite each other in a Frame can either be Complementary or non-Complementary.

In general, two numbers are complementary when they add up to a whole of some kind. For example, in geometry, two angles are said to be complementary when they add up to 90°. Also in Arithmetic, two numbers are said to be Complementary when they add-up to 10.

In the Frame 3 of a Perfect Afamatrix Magic Square Grid which is the outermost frame of the 7 X 7 grid or level 3 Afamatrix, all the digits from 1 to 9 are used three times (3X) each.

	1	**9**	8	**4**	3	→	25
	9				**1**	→	10
	6				4	→	10
	2				**8**	→	10
	7	**1**	2	**6**	9	→	25
↙	↓	↓	↓	↓	↓	↘	
10	25	10	10	10	25	10	

Types of Magic Square Grid Frames

Magic Square Grid Frames can be classified according to the type of Magic Square Grid it forms. Therefore, since there are two major types of Magic Square Grids namely Magic Square Grids of Odd order and Magic Square Grids of Even Order, Magic Square Grid Frames can also be classified into two namely;

1. Frames of Odd Order.
2. Frames of Even Order.

As the name implies, the Frames of Odd Order are the Frames of Magic Square Grids of Odd number Order while Frames of Even Order are Frames of Magic Square Grids of Even number Order.

In addition, depending on the arrangement of numbers in the Squares, Magic Square Grid Frames can be further divided into two (2) sets of subcategories, namely;

(i). Complementary, and

(ii). Non-Complementary

The numbers in the Squares opposite to each other in the Magic Square Grid Frames can either be complementary on non-Complementary;

Also, resultant sum of the numbers in the Top, Down, Right and Left Sides of the Magic Square Grid Frames can either be Magical or No-Magical.

(i). Magical, or
(ii). Non-Magical
Therefore, Eight Categories of Magic Square Grid Frames can be deciphered from the above major categories and sub-categories namely.

1) Complementary Magical Frames of Odd Order
2) Complementary Non-Magical Frames of Odd Order
3) Non-Complementary Magical Frames of Odd Order
4) Non-Complementary Non-Magical Frames of Odd Order

5) Complementary Magical Frames of Even Order
6) Complementary Non-Magical Frames of Even Order
7) Non-Complementary Magical Frames of Even Order
8) Non-Complementary Non-Magical Frames of Even Order

i. Complementary Magical Frames

In Complementary Magical Frames, the numbers in the squares opposite each other in the Frames are complementary, I.e. they add up to a consistent whole number (eg 10) or the result of arithmetic operation on the whole number.

The arrows in the Magic Square Frame below indicate complementary numbers, i.e. each pair of the numbers in the opposite squares adds up to 10.

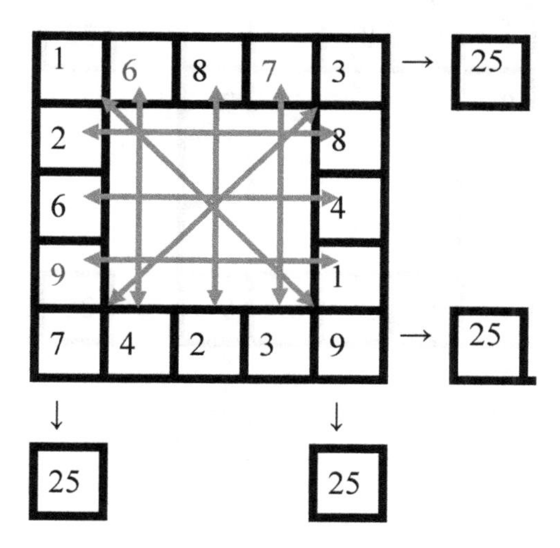

ii. Complementary Non-Magical Frames

In Complementary Non-Magical Frames, the numbers in the squares opposite each other in the Frames are complementary, I.e. they add up to a whole number (eg 10) or the result of arithmetic operation on the whole number but the sum of the total numbers on each side of the Frame are not Magical i.e Non-consistent . Such Frames needs other Frames with suitable inconsistency either internally or externally in order to form a consistent Magic Sum.

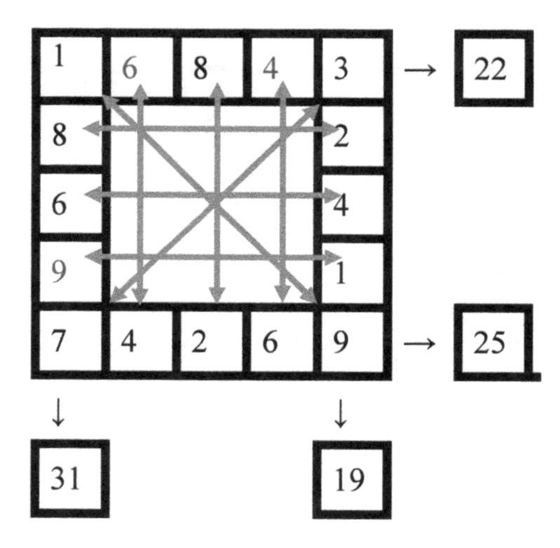

iii. Non-Complementary Magical Frames

In Non-Complementary Magical Frames, the numbers in the squares opposite each other in the Frames are not complementary, I.e. they do add up to a consistent whole number (eg 10) or the result of arithmetic operation on the whole number but the sum of the total numbers on each side of the Frame are Magical.

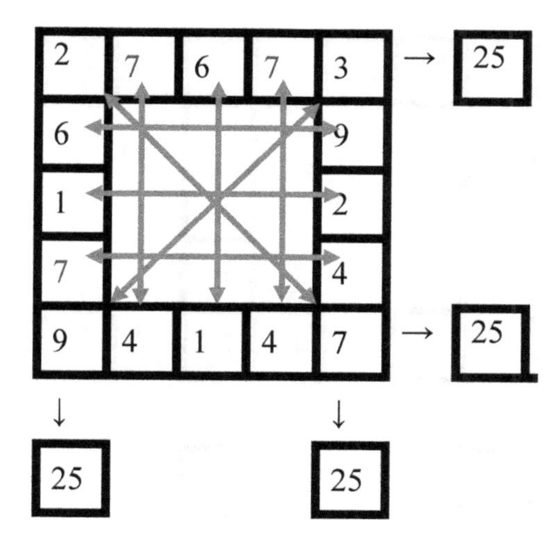

iv. Non-Complementary Non-Magical Frames

In Non-Complementary Non-Magical Frames, the numbers in the squares opposite each other in the Frames are not complementary, I.e. they do add up to a consistent whole number (e.g. 10) or the result of arithmetic operation on the whole number, and the sum of the total numbers on each side of the Frame are not Magical i.e. Non-consistent . Such Frames also needs other Frames with suitable inconsistency in both the Complementary nature of the numbers and the Magical Sum either internally or externally in order to form a consistent Magic Sum.

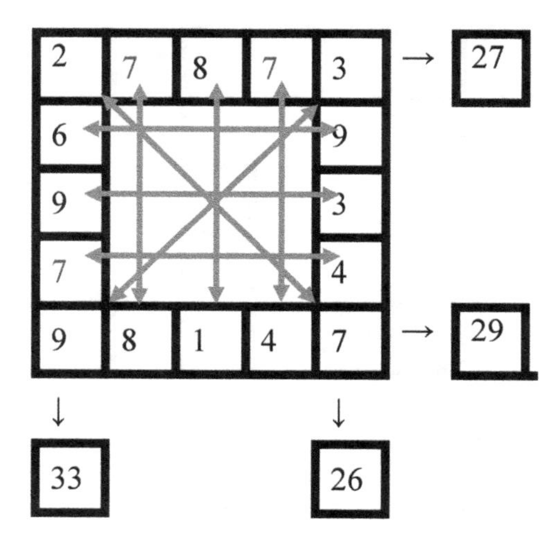

Numbering of Frames/Levels in Magic Square Grids

As already mentioned, Magic Square Grids are made up of Frames, the number of frames of which is dependent upon the size of the Magic Square Grid. The number of Frames a Magic Square Grid can have is only limited by the size of the Magic Square Grid. In this section we will take a look at how the Frames of Odd Order and Even Order Magic Square Grids are numbered.

Frames of Odd Order

For odd Ordered Magic Square Grids, numbering of the Frames starts from the innermost square, which could be regarded as Centre Square 0, and followed by Frame 1, Frame 2, Frame 3, Frame 4 and so on, outwards.

Centre Square 0

Frame 1

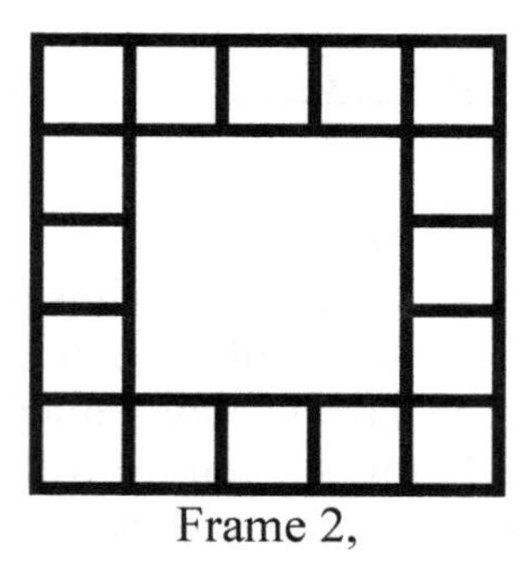

Frame 2,

Grid showing Frame 0 to Frame 11 of Odd Ordered Magic Square Grids.

NUMBERING OF ODD ORDER FRAMES IN MAGIC SQUARE GRIDS			
FRAME NUMBER	**ORDER**	**LEVEL**	**GRID SIZE (n^2)**
0	1	0	1 * 1
1	3	1	3 * 3
2	5	2	5 * 5
3	7	3	7 * 7
4	9	4	9 * 9
?	?	?	?

Frames of Even Order

For Even numbered Magic Square Grids, numbering of the Frames starts from Frame 0, Frame 1, Frame 2, Frame 3, and so on, out wards. Numbers in Frame 0 can either be Complementary or Non-Complementary but they are always Non-Magical.

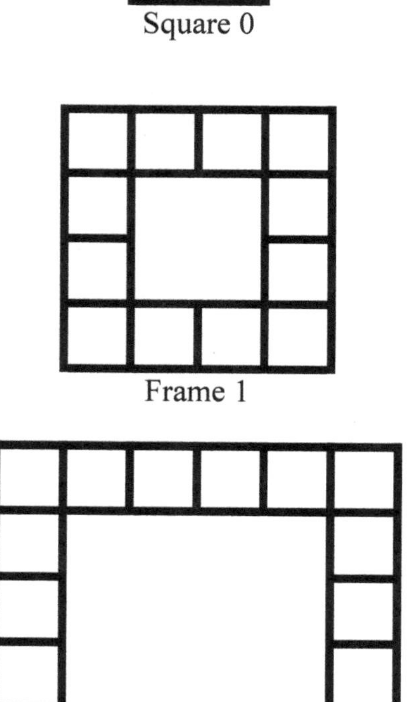

Square 0

Frame 1

Frame 2

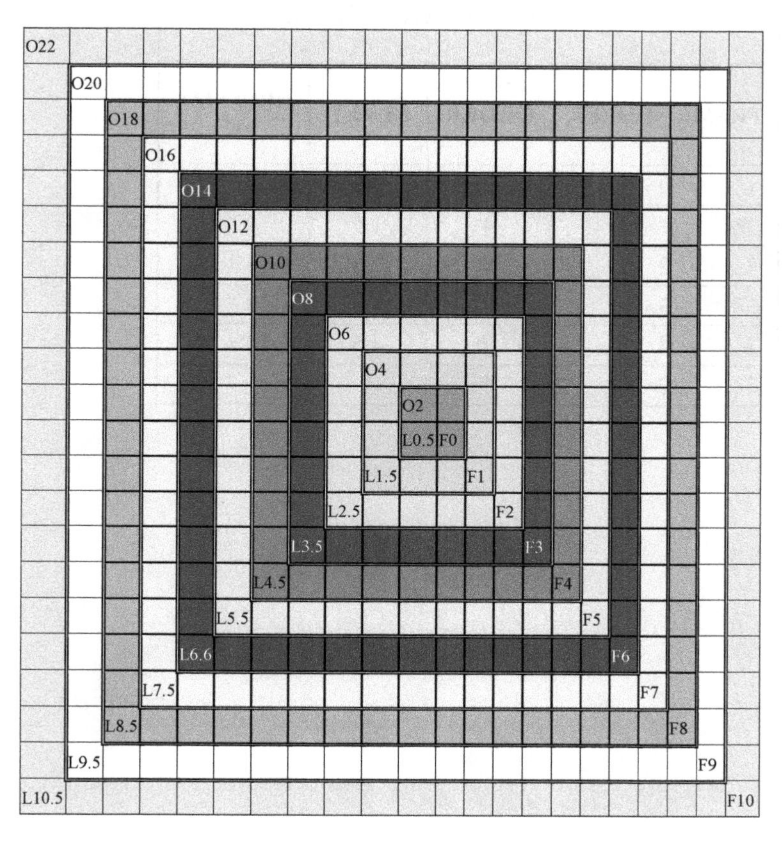

Grid showing Frame 0 to Frame 10 of Even Ordered Magic Square Grids.

NUMBERING OF EVEN ORDER FRAMES IN MAGIC SQUARE GRIDS			
FRAME NUMBER	**ORDER**	**LEVEL**	**GRID SIZE (n^2)**
0	2	0.5	2 * 2
1	4	1.5	4 * 4
2	6	2.5	6 * 6
3	8	3.5	8 * 8
4	10	4.5	10 * 10
?	?	?	?

Sub-components of Magic Square Grid Frames

The Magic Square Grid Frames are made up of two sub Frame Sub components. The X-Frame Subcomponent and the Plus Frame Subcomponent.

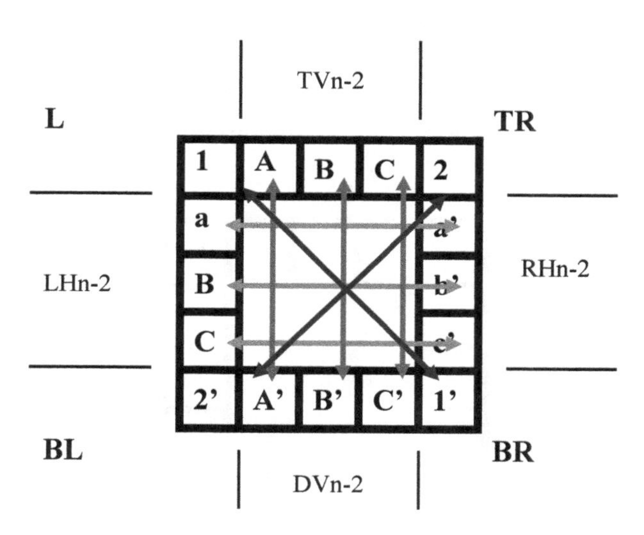

The X-Frame Subcomponent

A very important Subcomponent of the Frame considered in determining the variations of Magic Square Grids is the X-Frame Subcomponent.

Four squares make up an important Subcomponent of every Magic Square Grid Frame; these are the Top Leftmost Square (TL), also represented with the number 1. The Top Rightmost Square (TR) also represented with number 2. The Bottom Leftmost Square (BL), also represented with 2' (2 Compliment) and the Bottom Rightmost Square (BR) also represented with 1' (1 Compliment).

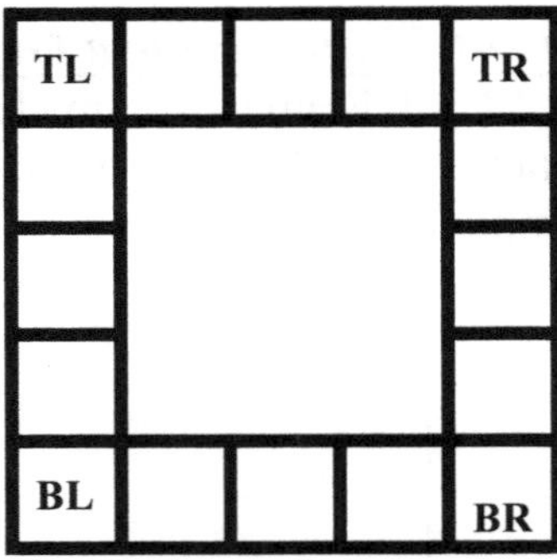

Representation of the X-Frame Subcomponents.

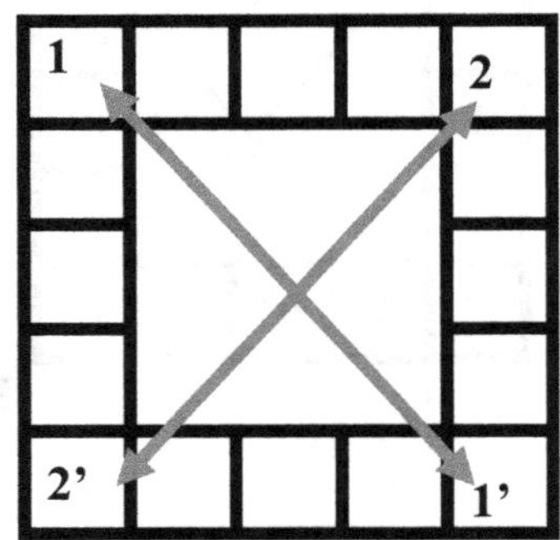

A representation of the Complementary nature of X-Frame Subcomponent numbers.

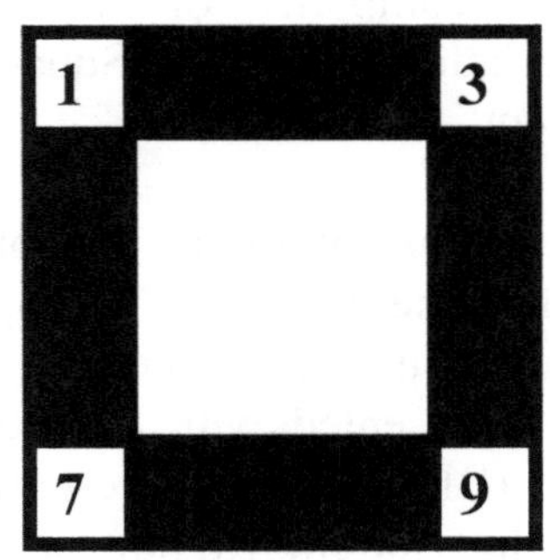

An example showing a Complementary X-Frame Subcomponent numbers.

Methods of Generating the X-Frame Subcomponents

In the Magic Square Grid Tables shown at the beginning of this Chapter, all the number sequences needed for the Magic Square Grid cells are generated simultaneously, i.e. 9 numbers for the 9 Cells of the Order 3 Magic Square Grid. In the case of Order 4, 4*4 = 20 cells will be involved, while in the case of Order 5, 5*5=25 cells will be involved. Considering the number of entities involved as already mentioned, generating number sequences for Magic Square Grids (either Incrementally or by Permutation) will be very difficult. As a result, in order to easily eliminate the invalid number sequences, numbers have to be generated according to the sub-divisions of the Magic Square Grid Components and Subcomponents.

In the case of the X-Frame Subcomponent of a Magic Square Grid, number sequences can be generated by any of the following four (4) ways;

1. Incrementing four (4) pair of numbers from the lowest given number to the highest given number in the given range of numbers.

2. Incrementing two (2) pair of numbers for one side of the X-Frame Subcomponent from the lowest given number to the highest given number in the range while filling the opposite cells with the complimentary numbers.

3. Permutation of the given Number Range and Four (4) which serves as the Chosen number range.

4. Permutation of the given Number Range and two (2) which serves as the Chosen number range, while filling the opposite cells with the complimentary numbers.

Consider the Basic Magic Square Grid Parameters Table below, you will notice that more parameters has been added; The "Lowest X-Frame Subcomponent Number" and "Highest X-Frame Subcomponent number". As specified in the table below, number repeat is allowed, and the maximum number quantity is 4, therefore the lowest X-Frame Sub component Number sequence will be 1111 while the Maximum will be 9999

BASIC MAGIC SQUARE GRID PARAMETERS TABLE		
S/N	**PARAMETERS**	**VALUE**
1	**Order / Level**	3 / 1
2	**Magic Sum**	15
3	**Type of Frame**	Non-Complementary Magical
4	**Type of Number Range**	Exclusive
5	**Number Range**	1 to 9
6	**RepeatableNumbers**	Yes
7	**Maximum Repeat Quantity**	4
8	**Number Range Exceptions**	Nil
9	**Exceptions Quantity**	Nil

X-Frame Subcomponent Table

S/N	Top Left	Top Right	Bottom Left	Bottom Right
	TL	TR	BL	BR
1	1	1	1	1
2	1	1	1	2
3	1	1	1	3
4	1	1	1	4
5	1	1	1	5
6	1	1	1	6
7	1	1	1	7
8	1	1	1	8
\ \ \ \		\ \ \ \ \ \		/ / / / / / /
\ \ \ \		\ \ \ \ \ \		/ / / / / / /
362873	9	9	9	2
362874	9	9	9	3
362875	9	9	9	4
362876	9	9	9	5
362877	9	9	9	6
362878	9	9	9	7
362879	9	9	9	8
362880	9	9	9	9

If number repeat is not allowed, then the maximum number quantity to be used in each set is 1. Therefore the lowest X-Frame Subcomponent Number Sequence will be 1234 while the highest will be 9876.

BASIC MAGIC SQUARE GRID PARAMETERS TABLE		
S/N	**PARAMETERS**	**VALUE**
1	**Order / Level**	3 / 1
2	**Magic Sum**	15
3	**Type of Frame**	Non-Complementary Magical
4	**Type of Number Range**	Exclusive
5	**Number Range**	1 to 9
6	**RepeatableNumbers**	No
7	**Maximum Repeat Quantity**	1
8	**Number Range Exceptions**	Nil
9	**Exceptions Quantity**	Nil

X-Frame Subcomponent Table

S/N	**Top Left**	**Top Right**	**Bottom Left**	**Bottom Right**
	TL	**TR**	**BL**	**BR**
1	1	2	3	4
2	1	2	3	5
3	1	2	3	6
4	1	2	3	7
5	1	2	3	8
6	1	2	3	9
7	1	2	4	3
8	1	2	4	5
...		...	...	...
...		...	...	...
...		...	...	...
362873	9	8	6	8
362874	9	8	6	9
362875	9	8	7	1
362876	9	8	7	2

362877	9	8	7	3
362878	9	8	7	4
362879	9	8	7	5
362880	9	8	7	6

The above two examples on Number Repeat considers that the numbers in the Magic Square Grid Frame are non-complementary, In the next two examples, we will consider a situation where the numbers are complementary, both for a Number repeat and a Non-number Repeat condition.

For a number repeat condition, where a complement of 10 is considered for example, the lowest and highest X-Frame Subcomponent numbers will be 1199 and 9911 respectively. This is shown in the basic Magic Square Grid Parameters Table and X-Frame Subcomponent Table below.

BASIC MAGIC SQUARE GRID PARAMETERS TABLE		
S/N	**PARAMETERS**	**VALUE**
1	**Order / Level**	3 / 1
2	**Magic Sum**	15
3	**Type of Frame**	Complementary Magical
4	**Type of Number Range**	Exclusive
5	**Number Range**	1 to 9
6	**Repeatable Numbers**	No
7	**Maximum Repeat Quantity**	1
8	**Number Range Exceptions**	Nil
9	**Exceptions Quantity**	Nil

X-Frame Subcomponent Table

S/N	**Top Left**	**Top Right**	**Bottom Left**	**Bottom Right**
	TL	**TR**	**BL**	**BR**
1	1	1	9	9
2	1	2	9	8
3	1	3	9	7
4	1	4	9	6
5	1	5	9	5
6	1	6	9	4
7	1	7	9	3
8	1	8	9	2
/ / / / .	\ \ \ \		/ / / / / / /	\ \ \ \ \ \ \
/ / / / .	\ \ \ \		/ / / / / / /	\ \ \ \ \ \ \
/ / / / .	\ \ \ \		/ / / / / / /	\ \ \ \ \ \ \
/ / / / .	\ \ \ \		/ / / / / / /	\ \ \ \ \ \ \
/ / / / .	\ \ \ \		/ / / / / / /	\ \ \ \ \ \ \
362873	9	2	1	8
362874	9	3	1	7
362875	9	4	1	6
362876	9	5	1	5

362877	9	6	1	4
362878	9	7	1	3
362879	9	8	1	2
362880	9	9	1	1

Also for a situation where number repeat is not allowed and complements of 10 are considered for example, then the lowest X-Frame Subcomponent numbers will be 1298 and 9812 respectively. This is also shown in the basic Magic Square Grid Parameters Table and X-Frame Subcomponent Table below.

BASIC MAGIC SQUARE GRID PARAMETERS TABLE		
S/N	**PARAMETERS**	**VALUE**
1	**Order / Level**	3 / 1
2	**Magic Sum**	15
3	**Type of Frame**	Complementary Magical
4	**Type of Number Range**	Exclusive
5	**Number Range**	1 to 9
6	**Repeatable Numbers**	N
7	**Maximum Repeat Quantity**	1
8	**Number Range Ex Ceptions**	Nil
9	**Exceptions Quantity**	Nil

X-Frame Subcomponent Table

S/N	Top Left	Top Right	Bottom Left	Bottom Right
	TL	TR	BL	BR
1	1	2	9	8
2	1	3	9	7
3	1	4	9	6
4	1	6	9	4
5	1	7	9	3
6	1	8	9	2
7	2	1	8	9
8	2	3	8	7
362873	8	3	2	7
362874	8	1	2	9
362875	9	2	1	8
362876	9	3	1	7
362877	9	4	1	6
362878	9	6	1	4
362879	9	7	1	3
362880	9	8	1	2

MORE DETAILS ON THE X-FRAME SUBCOMPONENT OF AN AFAMATRIX MAGIC SQUARE GRID

As already explained in the Previous Chapter on Afam's Method of Constructing Magic Square Grids, the X-Frame Subcomponent is a very important aspect of every magic Square Grid.

The X-Frame Subcomponent number selections are the group of numbers which are selected and arranged in the X-Frame Subcomponent squares in such a way as to create a complementary balance within the numbers while taking cognizance of the total numbers needed to make up a Magic Sum. Each group of numbers represents a unique Magic Square Frame position. The selections represent the various positions of the Magic Square Frame.

The set of X-Frame Subcomponent numbers are arranged in such a way as to create balance in the frame bearing in mind the complementary nature of the numbers, i.e. the pair of numbers that adds up to 10. For instance, if I choose the number 1 for starting point 1, and number 2 for starting point 2, then I have to complete the remaining starting points 3 and 4 with the numbers 9 and 8 respectively in other to create a complimentary balance.

The table below shows the position one (1) number selections starting from 1,1,9,9 to 5,5,5,5. Note that the selections starting from 1 stopped after 1,5,9,5, this is because if we should go ahead and list 1,6,9,4, and 1,7,9,3 and 1,8,9,2 and 1,9,9,1, we would have succeeded in repeating items '1', '2', '3' and 4 because they only depicts a change in frame position. As a result, we need to stop at 1,5,9,5, which is the middle number selection for the selections starting from 1. Likewise, the selections that started from 2 skipped 2,1,8,9 which is the would have been first selection because it is only a repetition of 1,2,9,8 which is item number 2 on the table, and simply a different frame position. The skipping continued for the selections starting from 3 and 4 till 5. Also note that the selections stopped at 5,5,5,5, which is the middle number selection, any selection beyond this is a repetition of what have been selected, it only represents a different frame position. For instance, if we should continue listing 6,1,4,9, we would have succeeded in listing a different frame position for 1,4,6,9, which is, item 4 on the table.

For instance, the Directional Method example shown below in the next chapter was illustrated using the level one 2,4,8,6, X-Frame Subcomponent. However, there are 14 other X-Frame Subcomponent number selections. In the latter sections and chapters, we will see the application of the number selections in determining the different frame positions and likewise different variations of Magic Square grids. This is fully explained in the section below titled "The X-Frame Subcomponent Groups". Frame Positioning buttresses the relationship between the X-Frame Sub component and the different Frame Positions.

The X-Frame Subcomponent Number Selections				
S/N	TOP LEFT (TL) 1	TOP RIGHT (TR) 2	BOTTOM RIGHT (BR) 1'	BOTTOM LEFT (BL) 2'
1	1	1	9	9
2	1	2	9	8
3	1	3	9	7
4	1	4	9	6
5	1	5	9	5
6	2	2	8	8
7	2	3	8	7
8	2	4	8	6
9	2	5	8	5
10	3	3	7	7
11	3	4	7	6
12	3	5	7	5
13	4	4	6	6
14	4	5	6	5
15	5	5	5	5

FRAME POSITIONING IN AFAMATRIX MAGIC SQUARE GRIDS USING THE X-FRAME SUB COMPONENT

The importance of the X-Frame Subcomponent part of Magic Square Grids cannot be fully understood without throwing more light into Frame Positioning, which is simply the different ways of rotating a **Number Sequence** in a Magic Square Grid Frame in other to form another Magic Square Grid Frame Position. A Number Sequence can be rotated in order to form another Number Sequence. Both the clockwise and anticlockwise rotated Number Sequences of every Magic Square Grid Frame has four (4) positions each making it a total of eight (8) Frame Positions for every pair of Number Sequence. This is irrespective of the order or size of the Magic Square Grid. However, some pair of Number Sequences do not necessarily form Eight (8) unique Frame Positions after rotation or reflection. Such Number Sequences can only have Six (6), Four (4) or One (1) unique Frame Position depending on the Complementary and Number Repeat specification of the Magic Square Grid Number Range. Once a Frame has been Positioned using the X-Frame Subcomponents Number Sequences, the Plus Component Compositions or the Permuted Partitions can then be filled-in.

The illustrations shown below uses the Order 3 Magic Square Grid to describe the eight positions which is achieved by simply reflecting or rotating a Magic Square Grid Frame to different positions.

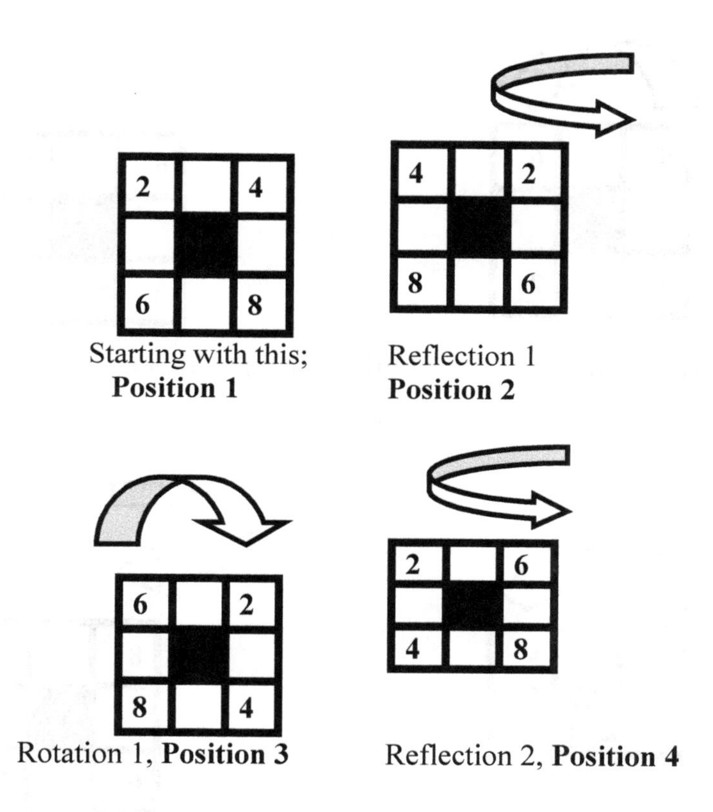

Starting with this; **Position 1**

Reflection 1 **Position 2**

Rotation 1, **Position 3**

Reflection 2, **Position 4**

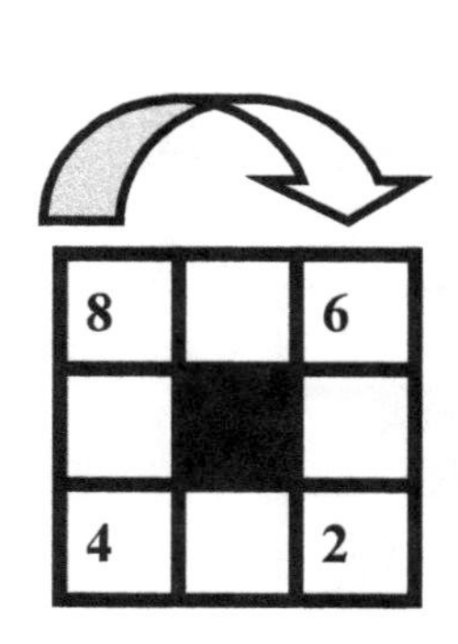

Rotation 2, **Position 5**

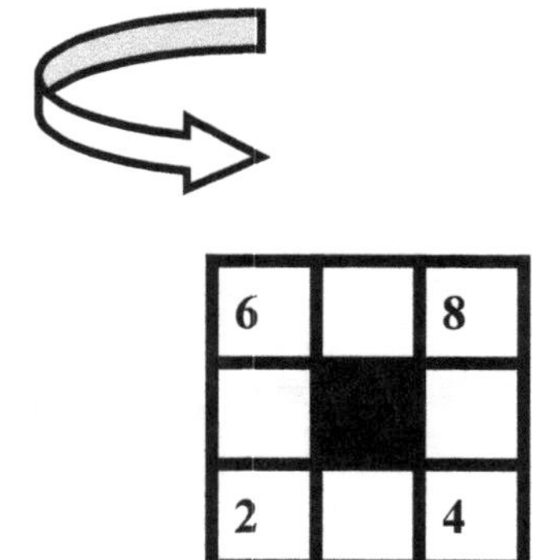

Reflection 3, **Position 6**

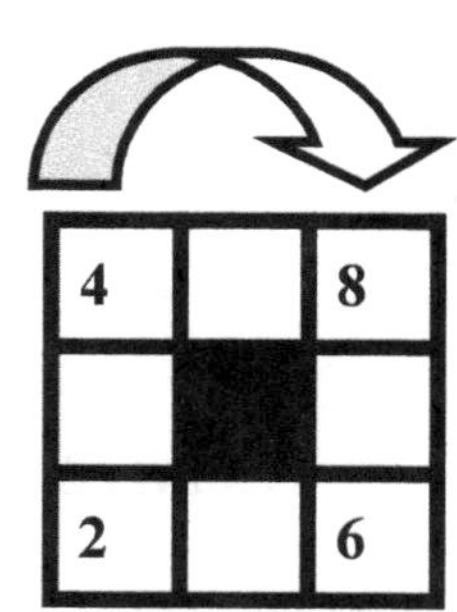

Rotation 3, **Position 7**

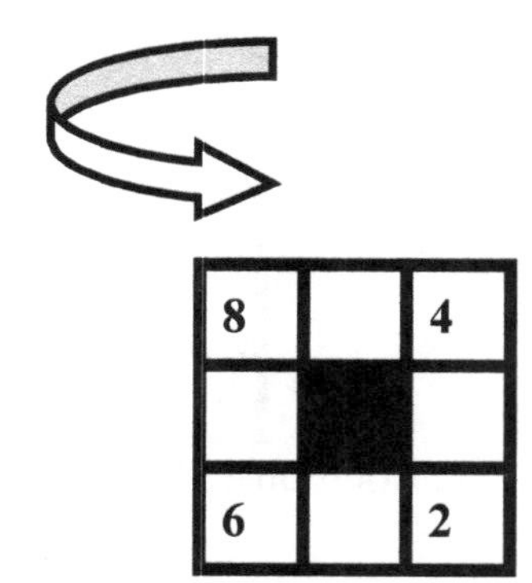

Reflection 4,**Position 8**

The "Directional Method" of Determining Frame Positions

In this section, I want to introduce to you a method for easily determining the Frame Positions of Magic Square Grid Frames, Since the positions play a very important role in determining the different variations of Magic Square Grids. I called it the **"Directional Method"** of determining Frame Positions. The method entails filling the squares with a **Number Sequence** starting from each of the Four (4)diagonal-end squares of the X-Frame Subcomponent, then moving forward in a clockwise direction (D1), also starting from each of the Four (4)Diagonal-End Squares, then moving backwards in an anticlockwise direction (D2). Formations of the eight possible positions are illustrated in the figures shown below:

Illustrations of the X-Frame Subcomponents Directional Method.

A		B
D		C

Four Starting Points of the Forward Clockwise Direction

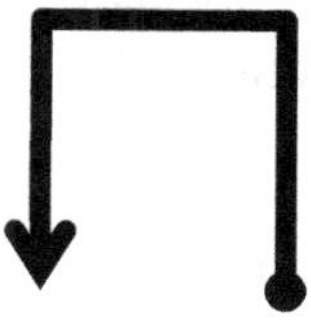

A		D
B		C

Four Starting Points of the Backwards Anti-Clockwise Direction

Take for instance the **2,4,8,6** Number Sequence, as shown below.

Position 1

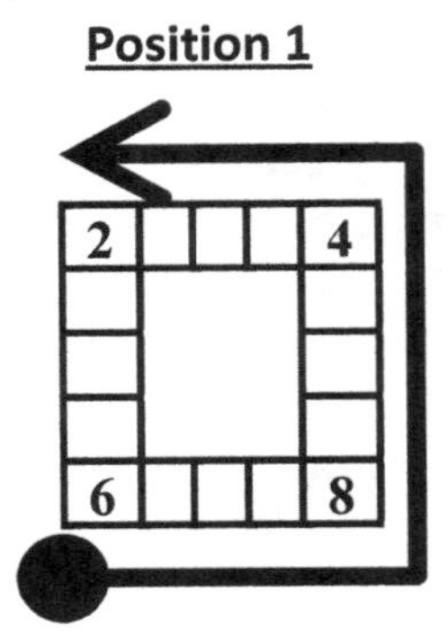

Starting Point 1 with forward Clockwise Direction (D1)

Position 2

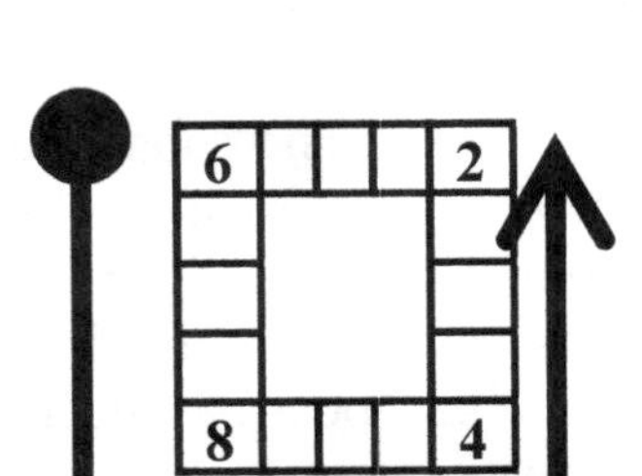

Starting Point 2 with forward Clockwise Direction (D1)

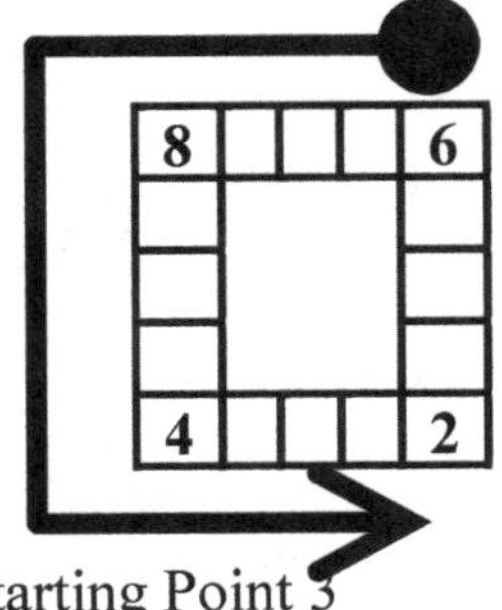

Starting Point 3 with forward Clockwise Direction (D1)

Position 4

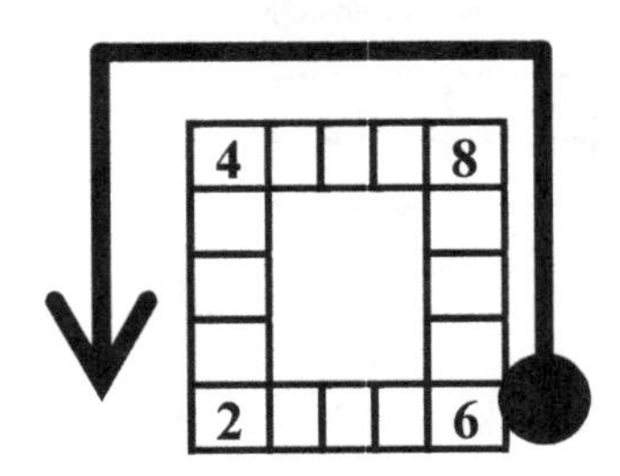

Starting Point 4 with forward Clockwise

Position 5

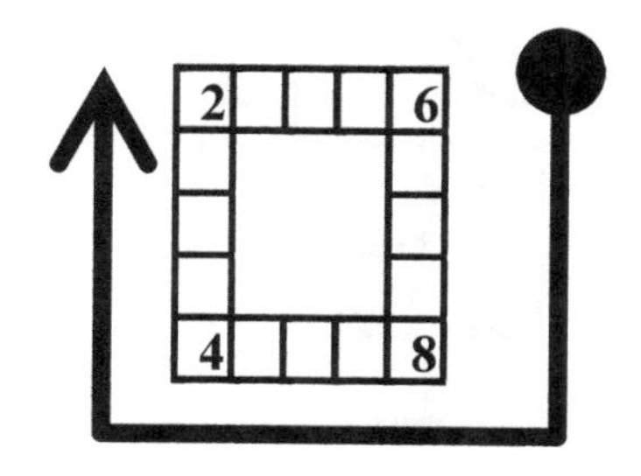

Starting Point 1 Starting Point 1 with Backwards Anti-Clockwise Direction (D2)

Position 6

Starting Point 2 Starting Point 1 with Backwards Anti-Clockwise Direction (D2)

Position 7

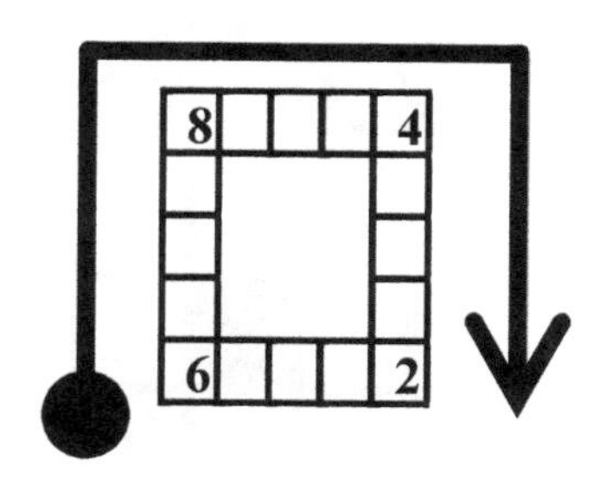

Starting Point 3 Starting Point 1 with Backwards Anti-Clockwise

Position 8 - - - -

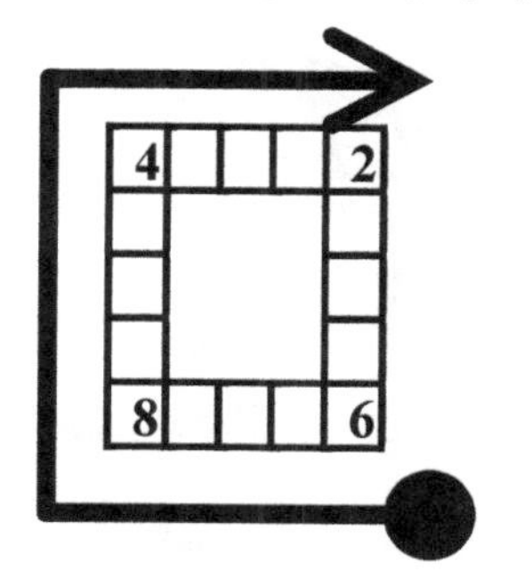

Starting Point 4 Starting Point 1 with Backwards Anti-Clockwise

AN EXAMPLE APPLICATION OF THE DIRECTIONAL METHOD IN DETERMINING THE EIGHT POSITIONS OF THE LEVEL ONE 2,4,6,8 X-COMPONENT SELECTION.

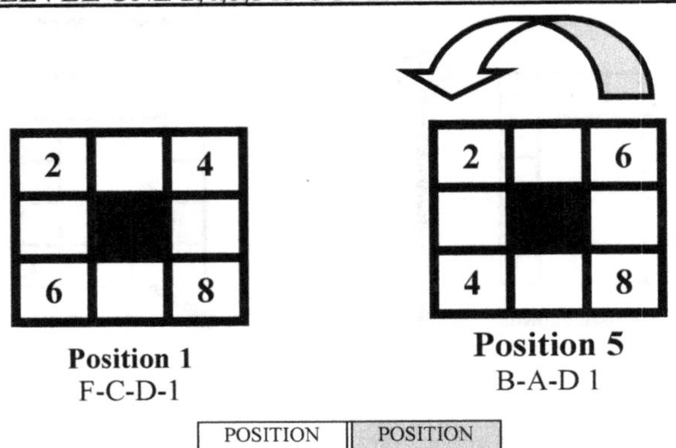

Position 1
F-C-D-1

Position 5
B-A-D 1

POSITION 1				POSITION 5			
TL	TR	BR	BL	TL	BL	BR	TR
2	4	8	6	2	4	8	6

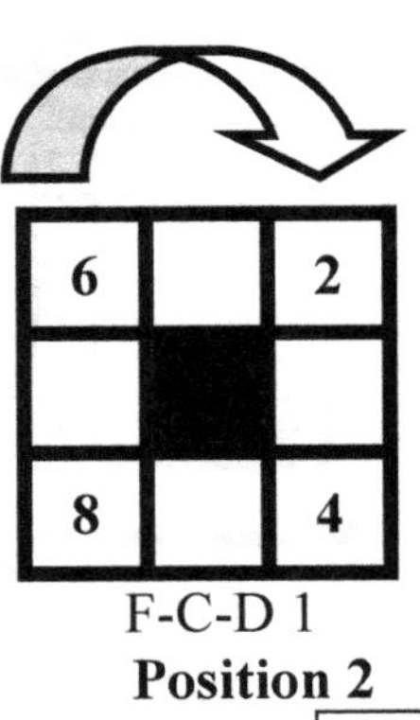

F-C-D 1
Position 2

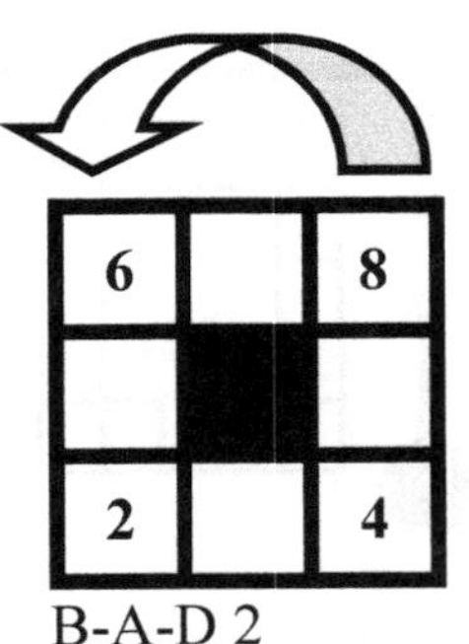

B-A-D 2
Position 6

POSITION 2				POSITION 6			
TL	TR	BR	BL	TL	BL	BR	TR
6	2	4	8	6	2	4	8

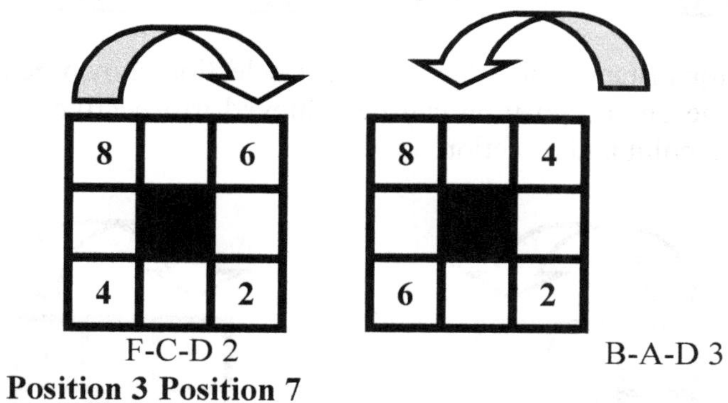

F-C-D 2 B-A-D 3

Position 3 Position 7

(D1) POSITION 3				(D2) POSITION 7			
TL	TR	BR	BL	TL	BL	BR	TR
1	2	1'	2'	1	2'	1'	2
8	6	2	4	8	6	2	4

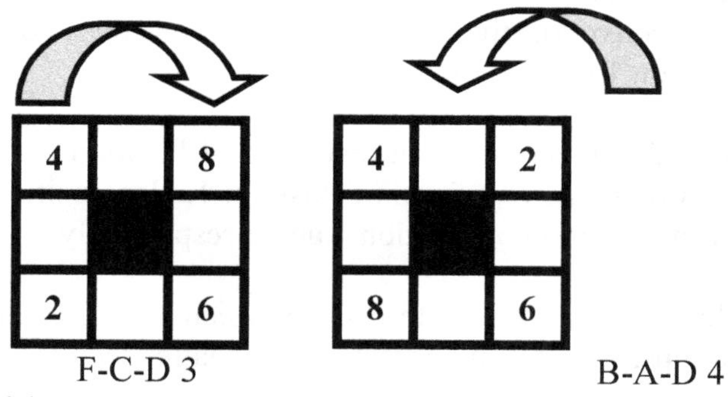

F-C-D 3 B-A-D 4

Position 4 Position 8

(D1) POSITION 4				(D2) POSITION 8			
TL	TR	BR	BL	TL	BL	BR	TR
1	2	1'	2'	1	2'	1'	2
4	8	6	2	4	8	6	2

UNIQUE POSITIONING IN X-COMPONENTS NUMBER SELECTIONS

Take for instance the 5,5,5,5 number selection shown below, only one unique position can be achieved irrespective of the starting point and direction.

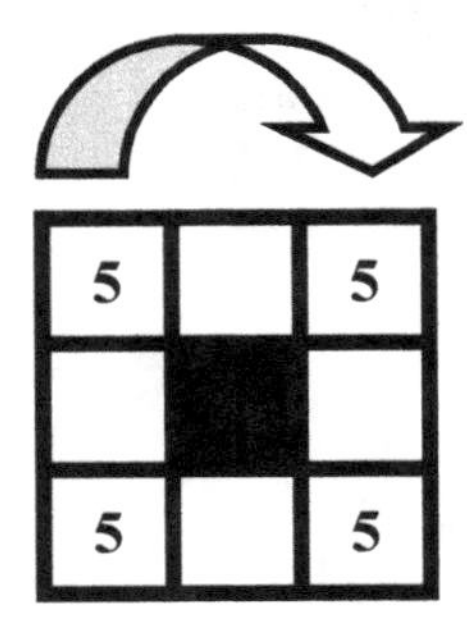

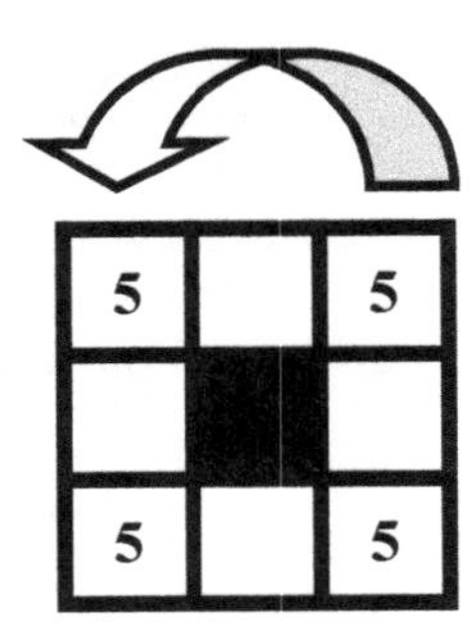

In another instance, the 2,8,8,2 number selections shown below gives only 4 unique positions instead of 8, irrespective of the starting position or direction. Only position 1 to 4 is unique.

Note that Position 5 is a repetition of Position 2, likewise position 6 is a repetition of Position 1, also position 7 and 8 is a repetition of position 4 and 3 respectively.

This explains the difference between the X-Component Groups, which will be treated in the latter section.

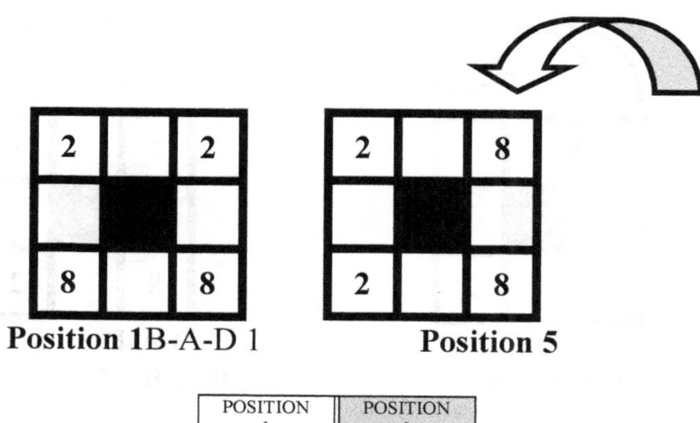

POSITION 1				POSITION 2			
TL	TR	BR	BL	TL	BL	BR	TR
2	8	8	2	2	8	8	2

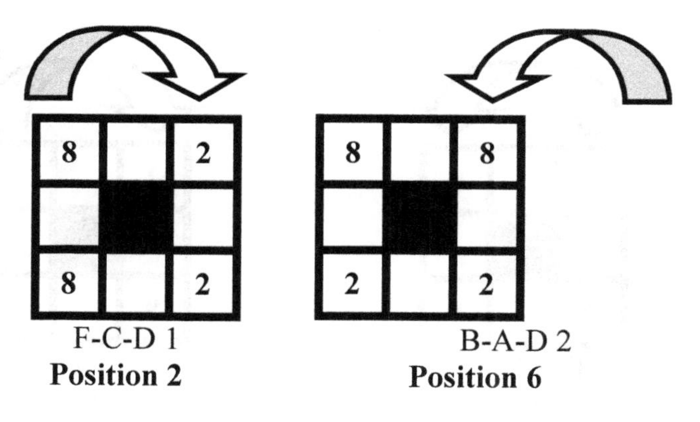

POSITION 3				POSITION 4			
TL	TR	BR	BL	TL	BL	BR	TR
2	2	8	8	2	2	8	8

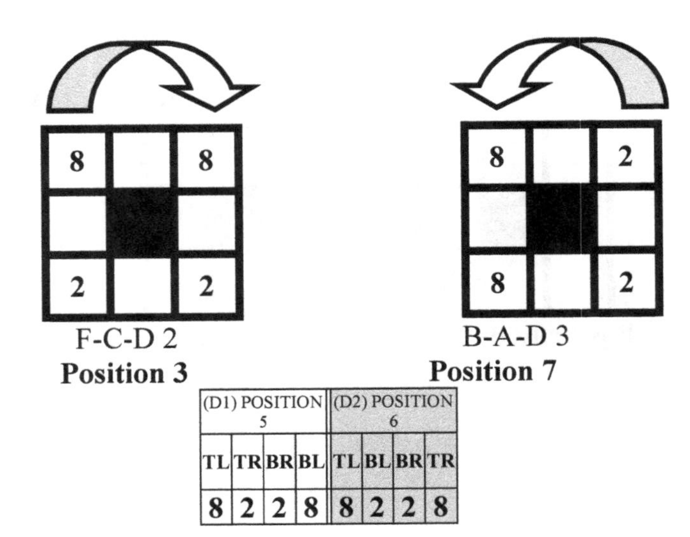

F-C-D 2
Position 3

B-A-D 3
Position 7

(D1) POSITION 5				(D2) POSITION 6			
TL	**TR**	**BR**	**BL**	**TL**	**BL**	**BR**	**TR**
8	**2**	**2**	**8**	**8**	**2**	**2**	**8**

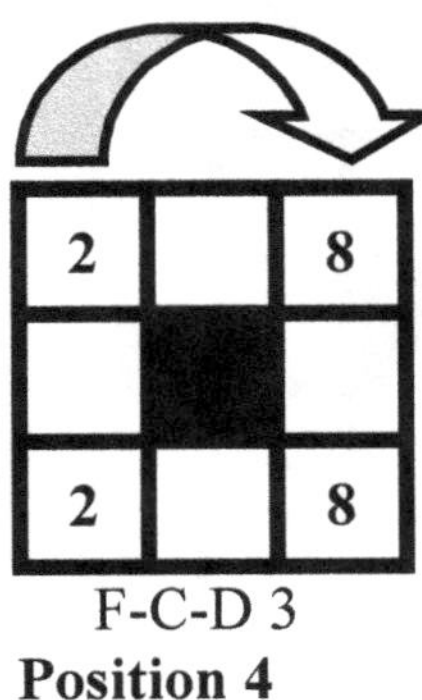

F-C-D 3
Position 4

B-A-D 4
Position 8

(D1) POSITION 7				(D2) POSITION 8			
TL	**TR**	**BR**	**BL**	**TL**	**BL**	**BR**	**TR**
8	**8**	**2**	**2**	**8**	**8**	**2**	**2**

DETERMINING THE UNIQUE X-COMPONENTS SELECTIONS

For us to have a better appreciation of how the unique X-Components can be determined, let us rearrange the Bi-Directional X-Component Number Selection Table shown above in the previous section such that the X-Component number selections flow in one direction.

For Instance, if the Backwards Anti-Clockwise Direction is chosen, then the columns in the Forward Clockwise Direction half part of the table has to be rearranged according to the Backwards Anti-Clockwise Direction. This means that the resultant number selection of the TL-TR-BR-BL will be written as TL-BL-BR-TR. Likewise, if the Forward Clockwise Direction is chosen, then the columns of the Backwards Anti-Clockwise Direction half part of the table have to be rearranged according to the Forward Clockwise Direction. This also means that the resultant number selection of the TL-BL-BR-TR will be written as TL-TR-BR-BL. The resultant table after this arrangement is the Single Directional X-Component Selection Table. This version of the table makes it easy for us to filter out the X-Components with repeated listing. The table will only contain Unique X-Component positions after the removal of the repeated X-Components.

As can be seen from the Single Direction X-Component Table shown below, the shaded X-Components in a selection are repetition of the un-shaded X-Components in the same selection.
The rearrangement from Bi-Directional X-Component table to Single Directional X-Component table can be seen below using the 2,4,6,8, X-Component as an example.

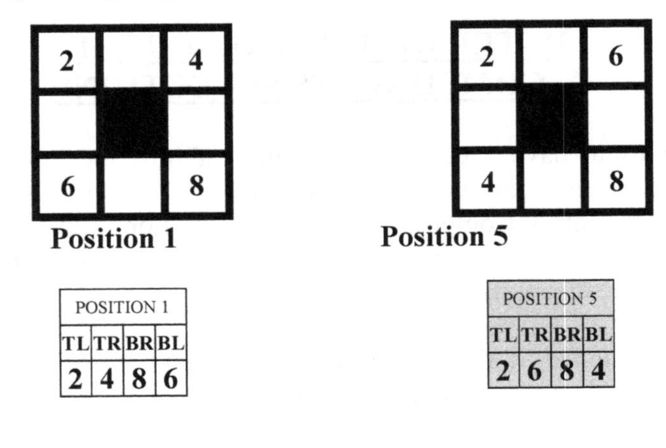

POSITION 1			
TL	TR	BR	BL
2	4	8	6

POSITION 5			
TL	TR	BR	BL
2	6	8	4

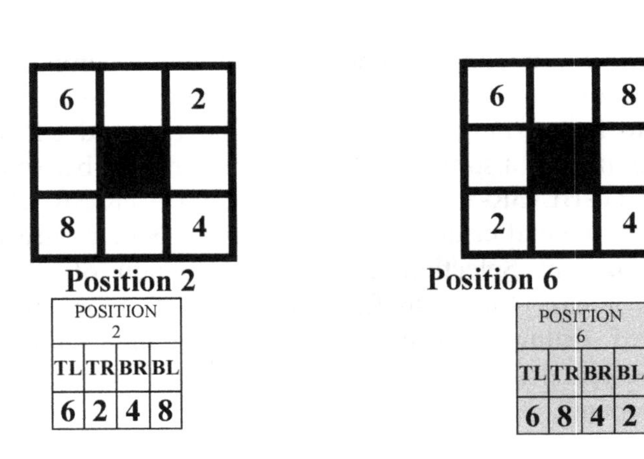

POSITION 2			
TL	TR	BR	BL
6	2	4	8

POSITION 6			
TL	TR	BR	BL
6	8	4	2

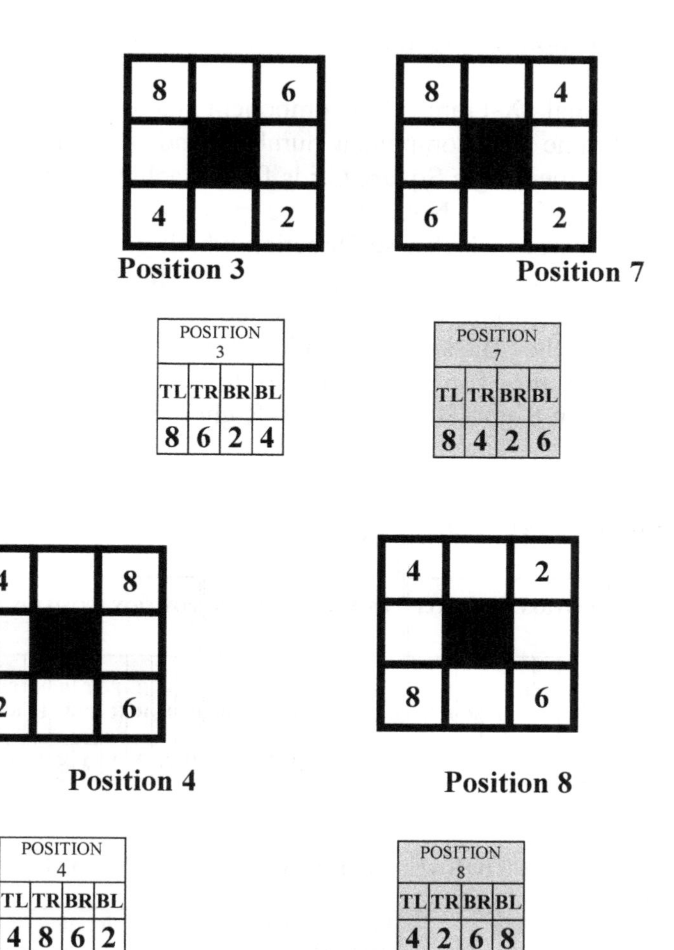

Position 3

Position 7

POSITION 3			
TL	TR	BR	BL
8	6	2	4

POSITION 7			
TL	TR	BR	BL
8	4	2	6

Position 4

Position 8

POSITION 4			
TL	TR	BR	BL
4	8	6	2

POSITION 8			
TL	TR	BR	BL
4	2	6	8

The Bi-Directional X-Frame Subcomponent Table

The Bi-Directional X-Frame Subcomponent table below shows the X-Frame Sub component numbers and how they are placed inside the Magic Square Grids for the achievement of the various positions. Note that one part of the table represents the Forward Clockwise Direction which goes from the Top Left (TL) to the Top Right (TR) through the Bottom Right (BR) to the Bottom Left (BL) while the 2nd part of the table represents the Backwards Anticlockwise Direction which goes from the Top Left (TL) to the Bottom Left through the Bottom Right (BR) to the Top Right (TR).

Forward Clockwise Direction (F-C-D) D1																**Backwards Anticlockwise Direction (B-A-D) D2**															
POSITION 1				**POSITION 2**				**POSITION 3**				**POSITION 4**				**POSITION 5**				**POSITION 6**				**POSITION 7**				**POSITION 8**			
TL	**TR**	**BR**	**BL**	**TL**	**TR**	**BR**	**BL**	**TL**	**TR**	**BR**	**BL**	**TL**	**TR**	**BR**	**BL**	**TL**	**BL**	**BR**	**TR**	**TL**	**BL**	**BR**	**TR**	**TL**	**BL**	**BR**	**TR**	**TL**	**BL**	**BR**	**TR**
2	4	8	6	6	2	4	8	8	6	2	4	4	8	6	2	2	4	8	6	6	2	4	8	8	6	2	4	4	8	6	2

COMPLETE BI-DIRECTIONAL X-FRAME SUBCOMPONENT SELECTION TABLE																																
	Forward Clockwise Direction (F-C-D) D1																**Backwards Anticlockwise Direction (B-A-D) D2**															
S/N	**POSITION 1**				**POSITION 2**				**POSITION 3**				**POSITION 4**				**POSITION 5**				**POSITION 6**				**POSITION 7**				**POSITION 8**			
	TL 1	**TR 2**	**BR 1'**	**BL 2'**	**TL 1**	**TR 2**	**BR 1'**	**BL 2'**	**TL 1**	**TR 2**	**BR 1'**	**BL 2'**	**TL 1**	**TR 2**	**BR 1'**	**BL 2'**	**TL 1**	**BL 2'**	**BR 1'**	**TR 2**	**TL 1**	**BL 2'**	**BR 1'**	**TR 2**	**TL 1**	**BL 2'**	**BR 1'**	**TR 2**	**TL 1**	**BL 2'**	**BR 1'**	**TR 2**
1	1	1	9	9	9	1	1	9	9	9	1	1	1	9	9	1	1	1	9	9	9	1	1	9	9	9	1	1	1	9	9	1
2	1	2	9	8	8	1	2	9	9	8	1	2	2	9	8	1	1	2	9	8	8	1	2	9	9	8	1	2	2	9	8	1
3	1	3	9	7	7	1	3	9	9	7	1	3	3	9	7	1	1	3	9	7	7	1	3	9	9	7	1	3	3	9	7	1

4	1	4	9	6	6	1	4	9	9	6	1	4	4	9	6	1	1	4	9	6	6	1	4	9	9	6	1	4	4	9	6	1
5	1	5	9	5	5	1	5	9	9	5	1	5	5	9	5	1	1	5	9	5	5	1	5	9	9	5	1	5	5	9	5	1
6	2	2	8	8	8	2	2	8	8	8	2	2	2	8	8	2	2	2	8	8	8	2	2	8	8	8	2	2	2	8	8	2
7	2	3	8	7	7	2	3	8	8	7	2	3	3	8	7	2	2	3	8	7	7	2	3	8	8	7	2	3	3	8	7	2
8	2	4	8	6	6	2	4	8	8	6	2	4	4	8	6	2	2	4	8	6	6	2	4	8	8	6	2	4	4	8	6	2
9	2	5	8	5	5	2	5	8	8	5	2	5	5	8	5	2	2	5	8	5	5	2	5	8	8	5	2	5	5	8	5	2
10	3	3	7	7	7	3	3	7	7	7	3	3	3	7	7	3	3	3	7	7	7	3	3	7	7	7	3	3	3	7	7	3
11	3	4	7	6	6	3	4	7	7	6	3	4	4	7	6	3	3	4	7	6	6	3	4	7	7	6	3	4	4	7	6	3
12	3	5	7	5	5	3	5	7	7	5	3	5	5	7	5	3	3	5	7	5	5	3	5	7	7	5	3	5	5	7	5	3
13	4	4	6	6	6	4	4	6	6	6	4	4	4	6	6	4	4	4	6	6	6	4	4	6	6	6	4	4	4	6	6	4
14	4	5	6	5	5	4	5	6	6	5	4	5	5	6	5	4	4	5	6	5	5	4	5	6	6	5	4	5	5	6	5	4
15	5	5	5	5	5	5	5	5	5	5	5	5	5	5	5	5	5	5	5	5	5	5	5	5	5	5	5	5	5	5	5	5

The Single-Directional X-Frame Subcomponent Table

What if the X-Frame Subcomponent numbers in the Magic Square Grid Frames are filled in the tables from one direction. The result will be a table that looks like the one below. Here, both part of the table starting from position 1 to position 8 flows in the Forward Clockwise Direction which goes from the Top Left (TL) to the Top Right (TR) through the Bottom Right (BR) to the Bottom Left (BL). Not that the filled-in numbers from position 5 to position 8 are no longer a repetition of the position 1 to position 4, like in the Bi-Directional Table. Instead the numbers are filled-in according to its appearance on the Frame.

Forward Clockwise Direction (F-C-D)
D1

POSITION 1				POSITION 2				POSITION 3				POSITION 4				POSITION 5				POSITION 6				POSITION 7				POSITION 8			
TL	TR	BR	BL	TL	TR	BR	BL	TL	TR	BR	BL	TL	TR	BR	BL	TL	TR	BR	BL	TL	TR	BR	BL	TL	TR	BR	BL	TL	TR	BR	BL
2	4	8	6	6	2	4	8	8	6	2	4	4	8	6	2	2	6	8	4	6	8	4	2	8	4	2	6	4	2	6	8

COMPLETE SINGLE-DIRECTIONAL X-FRAME SUBCOMPONENT SELECTION TABLE SHOWING THE INVALID POSITIONS
(Note the shaded repeated X-Frame Subcomponents)

S/N	POSITION 1				POSITION 2				POSITION 3				POSITION 4				POSITION 5				POSITION 6				POSITION 7				POSITION 8			
	TL 1	TR 2	BR 1'	BL 2'	TL 1	TR 2	BR 1'	BL 2'	TL 1	TR 2	BR 1'	BL 2'	TL 1	TR 2	BR 1'	BL 2'	TL 1	TR 2	BR 1'	BL 2'	TL 1	TR 2	BR 1'	BL 2'	TL 1	TR 2	BR 1'	BL 2'	TL 1	TR 2	BR 1'	BL 2'
1	1	1	9	9	9	1	1	9	9	9	1	1	1	9	9	1	1	9	9	1	9	9	1	1	9	1	1	9	1	1	9	9
2	1	2	9	8	8	1	2	9	9	8	1	2	2	9	8	1	1	8	9	2	8	9	2	1	9	2	1	8	2	1	8	9
3	1	3	9	7	7	1	3	9	9	7	1	3	3	9	7	1	1	7	9	3	7	9	3	1	9	3	1	7	3	1	7	9
4	1	4	9	6	6	1	4	9	9	6	1	4	4	9	6	1	1	6	9	4	6	9	4	1	9	4	1	6	4	1	6	9
5	1	5	9	5	5	1	5	9	9	5	1	5	5	9	5	1	1	5	9	5	5	9	5	1	9	5	1	5	5	1	5	9
6	2	2	8	8	8	2	2	8	8	8	2	2	2	8	8	2	2	8	8	2	8	8	2	2	8	2	2	8	2	2	8	8
7	2	3	8	7	7	2	3	8	8	7	2	3	3	8	7	2	2	7	8	3	7	8	3	2	8	3	2	7	3	2	7	8
8	2	4	8	6	6	2	4	8	8	6	2	4	4	8	6	2	2	6	8	4	6	8	4	2	8	4	2	6	4	2	6	8
9	2	5	8	5	5	2	5	8	8	5	2	5	5	8	5	2	2	5	8	5	5	8	5	2	8	5	2	5	5	2	5	8
10	3	3	7	7	7	3	3	7	7	7	3	3	3	7	7	3	3	7	7	3	7	7	3	3	7	3	3	7	3	3	7	7
11	3	4	7	6	6	3	4	7	7	6	3	4	4	7	6	3	3	6	7	4	6	7	4	3	7	4	3	6	4	3	6	7
12	3	5	7	5	5	3	5	7	7	5	3	5	5	7	5	3	3	5	7	5	5	7	5	3	7	5	3	5	5	3	5	7
13	4	4	6	6	6	4	4	6	6	6	4	4	4	6	6	4	4	6	6	4	6	6	4	4	6	4	4	6	4	4	6	6
14	4	5	6	5	5	4	5	6	6	5	4	5	5	6	5	4	4	5	6	5	5	6	5	4	6	5	4	5	5	4	5	6
15	5	5	5	5	5	5	5	5	5	5	5	5	5	5	5	5	5	5	5	5	5	5	5	5	5	5	5	5	5	5	5	5

REMOVAL OF THE INVALID POSITIONS IN THE SINGLE-DIRECTIONAL X-FRAME SUBCOMPONENT SELECTION TABLE

(Note that there are Eighty one (81) Possible unique Positions after the removal of the repeated X-Frame Subcomponents)

S/N	POSITION 1				POSITION 2				POSITION 3				POSITION 4				POSITION 5				POSITION 6				POSITION 7				POSITION 8				Unique Positions
	TL	TR	BR	BL	TL	TR	BR	BL	TL	TR	BR	BL	TL	TR	BR	BL	TL	TR	BR	BL	TL	TR	BR	BL	TL	TR	BR	BL	TL	TR	BR	BL	
1	1	1	9	9	9	1	1	9	9	9	1	1	1	9	9	1																	4
2	1	2	9	8	8	1	2	9	9	8	1	2	2	9	8	1	1	8	9	2	8	9	2	1	9	2	1	8	2	1	8	9	8
3	1	3	9	7	7	1	3	9	9	7	1	3	3	9	7	1	1	7	9	3	7	9	3	1	9	3	1	7	3	1	7	9	8
4	1	4	9	6	6	1	4	9	9	6	1	4	4	9	6	1	1	6	9	4	6	9	4	1	9	4	1	6	4	1	6	9	8
5	1	5	9	5	5	1	5	9	9	5	1	5	5	9	5	1																	4
6	2	2	8	8	8	2	2	8	8	8	2	2	2	8	8	2																	4
7	2	3	8	7	7	2	3	8	8	7	2	3	3	8	7	2	2	7	8	3	7	8	3	2	8	3	2	7	3	2	7	8	8
8	2	4	8	6	6	2	4	8	8	6	2	4	4	8	6	2	2	6	8	4	6	8	4	2	8	4	2	6	4	2	6	8	8
9	2	5	8	5	5	2	5	8	8	5	2	5	5	8	5	2																	4
10	3	3	7	7	7	3	3	7	7	7	3	3	3	7	7	3																	4
11	3	4	7	6	6	3	4	7	7	6	3	4	4	7	6	3	3	6	7	4	6	7	4	3	7	4	3	6	4	3	6	7	8
12	3	5	7	5	5	3	5	7	7	5	3	5	5	7	5	3																	4
13	4	4	6	6	6	4	4	6	6	6	4	4	4	6	6	4																	4
14	4	5	6	5	5	4	5	6	6	5	4	5	5	6	5	4																	4
15	5	5	5	5																													1

Number Sets and Sequences of X-Frame Sub Component Magic Square Grid Positions.

The total possible X-Frame Subcomponent Positions of a given Number Range of the Magic Square Grid is generated either Incrementally in groups of four (4) numbers, or by the permutation of the given Number Range using four (4) as the chosen number of elements (since there are four (4) X-Frame Subcomponent boxes). This depends on the Complementary and Number Repeat specification of the Magic Square Grid. Each Group of four Numbers represents a Frame Position.

The total Frame Positions that resulted from the Permutation can be further divided into unique **Number Sets**. A Number Set contains four (4) unique digits irrespective of the Sequence (ordering)of the digits. The Number Sets can be found using two methods. One method is by the Mathematical Combination of the given Number Range of the Magic Square Grid using four (4) as the Chosen number of Elements.

Understanding Combinations.

Before proceeding further, let us throw more light on Combinations. Combination is quite similar to Permutations but in Combinations, only unique number sets are needed. Combination will also be applied at the latter end of this book when we will be discussing about the Practical Application of Magic Square Grids in Generating Magic Square Grid Puzzles.

Unlike Permutation, Combination has to do with the number of ways of selecting ***r*** objects from ***n*** unlike objects, and it is written as:

$$\binom{n}{r} = {}^{n}C_{r} = \frac{n!}{r!\,(n-r)!}$$

Example

There are 10 Oranges in a bag numbered from 1 to 10. If three Oranges are selected at random. How many different ways are there for selecting the three Oranges?

$$^{10}C_3 = \frac{10!}{3!\,(10-3)!} = \frac{10 \times 9 \times 8}{3 \times 2 \times 1} = \underline{120}$$

In Permutations, the order is important while in Combination, the order is not important. In Permutation, we count **abc** as different from **bca**. But in combinations we are concerned only that **a, b,** and **c**have been selected. **abc** and **bca** are the same combination. The other method of finding the unique **Number Set s**in the permuted Positions of a Magic Square Grid Frame is by dividing the total number of permuted Frame positions of the number range by the number of permutations of any set of four numbers from the range depending on the number repeat specifications of the magic square grid. For a case where there is no number repeat, the number of permutations of any set of four numbers is twenty four (24). For a case where number repeat is two (2), three (3) and four (4), the number of permutations will be twelve (12), six (6) and one (1) respectively.
Each Number Set can also be divided into Unique **Number Sequences** while each Number Sequence can be rotated into unique Frame Positions, as described at the beginning of this section. The Possible number of generated Number Sets, Number Sequences and Frame Positions is different for a case where Number Repeat is allowed and a case where Number repeat is not allowed. They are also different for cases where the use of Complementary numbers are specified and cases where it is not.

For Instance, the 1-9 Number Range as shown at the beginning of the Chapter on Afam's Method of Constructing Magic Square Grids which have 362,880 Permutations (Since Number Repeat is not allowed) will contain 3024 X-Frame Subcomponent

Permutations or Frame Positions. This is found by permuting the given Number Range 1-9 and 4 as the Chosen Number Range. The 3024 Positions can be divided into 126 Unique Number Sets using two methods, one of which is Mathematical Combination. The other method is by dividing the total number of permuted Frame positions of the number range which is 3024 by the number of permutations of any set of four numbers which is equal to 24 since number repeat is not allowed. Both methods will give 126 number sets. Therefore, each Number Set has 24 unique positions. The 24 positions can be divided into 6 unique Number Sequences. Each Number Sequence can be rotated into 4 unique Positions. Three 3 out of the 6Number Sequences represents the Forward Clockwise Movement while the other 3 represents the Backwards Anticlockwise Movement. The 126 Unique Number Sets of the 1-9 Number Range are shown in the Tables below.

126 Unique Number Sets of the 1-9 Number Range

	TL	TR	BR	BL
1	1	2	3	4
2	1	2	3	5
3	1	2	3	6
4	1	2	3	7
5	1	2	3	8
6	1	2	3	9
7	1	2	4	5
8	1	2	4	6
9	1	2	4	7
10	1	2	4	8
11	1	2	4	9
12	1	2	5	6
13	1	2	5	7

	TL	TR	BR	BL
22	1	3	4	5
23	1	3	4	6
24	1	3	4	7
25	1	3	4	8
26	1	3	4	9
27	1	3	5	6
28	1	3	5	7
29	1	3	5	8
30	1	3	5	9
31	1	3	6	7
32	1	3	6	8
33	1	3	6	9
34	1	3	7	8

	TL	TR	BR	BL
43	1	4	6	9
44	1	4	7	8
45	1	4	7	9
46	1	4	8	9
47	1	5	6	7
48	1	5	6	8
49	1	5	6	9
50	1	5	7	8
51	1	5	7	9
52	1	5	8	9
53	1	6	7	8
54	1	6	7	9
55	1	6	8	9

	TL	TR	BR	BL
64	2	3	5	8
65	2	3	5	9
66	2	3	6	7
67	2	3	6	8
68	2	3	6	9
69	2	3	7	8
70	2	3	7	9
71	2	3	8	9
72	2	4	5	6
73	2	4	5	7
74	2	4	5	8
75	2	4	5	9
76	2	4	6	7

	TL	TR	BR	BL
85	2	5	7	8
86	2	5	7	9
87	2	5	8	9
88	2	6	7	8
89	2	6	7	9
90	2	6	8	9
91	2	7	8	9
92	3	4	5	6
93	3	4	5	7
94	3	4	5	8
95	3	4	5	9
96	3	4	6	7
97	3	4	6	8

	TL	TR	BR	BL
106	3	5	7	9
107	3	5	8	9
108	3	6	7	8
109	3	6	7	9
110	3	6	8	9
111	3	7	8	9
112	4	5	6	7
113	4	5	6	8
114	4	5	6	9
115	4	5	7	8
116	4	5	7	9
117	4	5	8	9
118	4	6	7	8

14	1	2	5	8	35	1	3	7	9	56	1	7	8	9	77	2	4	6	8	98	3	4	6	9	119	4	6	7	9
15	1	2	5	9	36	1	3	8	9	57	2	3	4	5	78	2	4	6	9	99	3	4	7	8	120	4	6	8	9
16	1	2	6	7	37	1	4	5	6	58	2	3	4	6	79	2	4	7	8	100	3	4	7	9	121	4	7	8	9
17	1	2	6	8	38	1	4	5	7	59	2	3	4	7	80	2	4	7	9	101	3	4	8	9	122	5	6	7	8
18	1	2	6	9	39	1	4	5	8	60	2	3	4	8	81	2	4	8	9	102	3	5	6	7	123	5	6	7	9
19	1	2	7	8	40	1	4	5	9	61	2	3	4	9	82	2	5	6	7	103	3	5	6	8	124	5	6	8	9
20	1	2	7	9	41	1	4	6	7	62	2	3	5	6	83	2	5	6	8	104	3	5	6	9	125	5	7	8	9
21	1	2	8	9	42	1	4	6	8	63	2	3	5	7	84	2	5	6	9	105	3	5	7	8	126	6	7	8	9

For instance, the 2,4,6,8 Number Set (S/N. 77 in the table above) have 24 Permutations. The 24 Permutations can be divided into 6 Number Sequences.

The 24 Permutations of the 2,4,6,8 Number Set.

	TL	TR	BR	BL		TL	TR	BR	BL		TL	TR	BR	BL
1	2	4	6	8	9	4	6	2	8	17	6	8	2	4
2	2	4	8	6	10	4	6	8	2	18	6	8	4	2
3	2	6	4	8	11	4	8	2	6	19	8	2	4	6
4	2	6	8	4	12	4	8	6	2	20	8	2	6	4
5	2	8	4	6	13	6	2	4	8	21	8	4	2	6
6	2	8	6	4	14	6	2	8	4	22	8	4	6	2
7	4	2	6	8	15	6	4	2	8	23	8	6	2	4
8	4	2	8	6	16	6	4	8	2	24	8	6	4	2

The Six (6) Number Sequences of the 2,4,6,8 Number Set.

	TL	TR	BR	BL	
1	2	4	6	8	C1
2	2	4	8	6	C2
3	2	6	4	8	C3
4	2	6	8	4	A2
5	2	8	4	6	A3
6	2	8	6	4	A1

Three (3) out of the Six (6) Number Sequences represents the Forward Clockwise Movement while the other three (3) represents the Backwards Anticlockwise Movement, I a single directional table arrangement. The 2,8,6,4 and 2,6,8,4 and 2,8,4,6 number sequences are the anti-clockwise movement of the 2,4, 6,8 and 2,4,8,6 and 2,6,4,8 number sequences respectively.

	TL	TR	BR	BL				TL	TR	BR	BL	
1	2	4	6	8	C1	→	6	2	8	6	4	A1
2	2	4	8	6	C2	→	4	2	6	8	4	A2
3	2	6	4	8	C3	→	5	2	8	4	6	A3

Note that there is no way any of the Number Sequences can be rotated clockwise or anticlockwise to form the same position as any other Number Sequence among the six (6).
Each Number Sequence among the six (6) can be rotated into four (4) Unique Positions, making a total of 24 Unique Positions, as shown in the tables below.

	TL	TR	BR	BL
C1	2	4	6	8
1	**2**	**4**	**6**	**8**
2	**8**	**2**	**4**	**6**
3	**6**	**8**	**2**	**4**
4	**4**	**6**	**8**	**2**

	TL	TR	BR	BL
C2	2	4	8	6
1	**2**	**4**	**8**	**6**
2	**6**	**2**	**4**	**8**
3	**8**	**6**	**2**	**4**
4	**4**	**8**	**6**	**2**

	TL	TR	BR	BL
C3	2	6	4	8
1	**2**	**6**	**4**	**8**
2	**8**	**2**	**6**	**4**
3	**4**	**8**	**2**	**6**
4	**6**	**4**	**8**	**2**

	TL	TR	BR	BL
A1	2	8	6	4
1	**2**	**8**	**6**	**4**
2	**4**	**2**	**8**	**6**
3	**6**	**4**	**2**	**8**
4	**8**	**6**	**4**	**2**

	TL	TR	BR	BL
A2	2	6	8	4
1	**2**	**6**	**8**	**4**
2	**4**	**2**	**6**	**8**
3	**8**	**4**	**2**	**6**
4	**6**	**8**	**4**	**2**

	TL	TR	BR	BL
A3	2	8	4	6
1	**2**	**8**	**4**	**6**
2	**6**	**2**	**8**	**4**
3	**4**	**6**	**2**	**8**
4	**8**	**4**	**6**	**2**

The Clockwise Number Sequence together with the corresponding Anti-Clockwise Number Sequence contains a **Number Sequence Pair** which forms eight (8) Unique Positions each.

	TL	TR	BR	BL
C1	2	4	6	8
A1	2	8	6	4
1	**2**	**4**	**6**	**8**
2	**8**	**2**	**4**	**6**
3	**6**	**8**	**2**	**4**
4	**4**	**6**	**8**	**2**

	TL	TR	BR	BL
C2	2	4	8	6
A2	2	6	8	4
1	**2**	**4**	**8**	**6**
2	**6**	**2**	**4**	**8**
3	**8**	**6**	**2**	**4**
4	**4**	**8**	**6**	**2**

	TL	TR	BR	BL
C3	2	6	4	8
A3	2	8	4	6
1	**2**	**6**	**4**	**8**
2	**8**	**2**	**6**	**4**
3	**4**	**8**	**2**	**6**
4	**6**	**4**	**8**	**2**

5	2	8	6	4
6	4	2	8	6
7	6	4	2	8
8	8	6	4	2

5	2	6	8	4
6	4	2	6	8
7	8	4	2	6
8	6	8	4	2

5	2	8	4	6
6	6	2	8	4
7	4	6	2	8
8	8	4	6	2

Use of Number Sets in Complementary Magic Square Grid Frames.

In a Complementary Magic Square Grid Frame, the numbers in the opposite squares are complementary to each other. Only 48 out of the 3024 Number sets can form a complementary frame.

	TL	TR	BR	BL	TL + BR	TR + BL
1	1	2	9	8	10	10
2	1	3	9	7	10	10
3	1	4	9	6	10	10
4	1	6	9	4	10	10
5	1	7	9	3	10	10
6	1	8	9	2	10	10
7	2	1	8	9	10	10
8	2	3	8	7	10	10
9	2	4	8	6	10	10
10	2	6	8	4	10	10
11	2	7	8	3	10	10
12	2	9	8	1	10	10
13	3	1	7	9	10	10
14	3	2	7	8	10	10
15	3	4	7	6	10	10
16	3	6	7	4	10	10
17	3	8	7	2	10	10
18	3	9	7	1	10	10
19	4	1	6	9	10	10
20	4	2	6	8	10	10

	TL	TR	BR	BL	TL + BR	TR + BL
25	6	1	4	9	10	10
26	6	2	4	8	10	10
27	6	3	4	7	10	10
28	6	7	4	3	10	10
29	6	8	4	2	10	10
30	6	9	4	1	10	10
31	7	1	3	9	10	10
32	7	2	3	8	10	10
33	7	4	3	6	10	10
34	7	6	3	4	10	10
35	7	8	3	2	10	10
36	7	9	3	1	10	10
37	8	1	2	9	10	10
38	8	3	2	7	10	10
39	8	4	2	6	10	10
40	8	6	2	4	10	10
41	8	7	2	3	10	10
42	8	9	2	1	10	10
43	9	2	1	8	10	10
44	9	3	1	7	10	10

21	**4**	**3**	**6**	**7**	10	10		45	**9**	**4**	**1**	**6**	10	10
22	**4**	**7**	**6**	**3**	10	10		46	**9**	**6**	**1**	**4**	10	10
23	**4**	**8**	**6**	**2**	10	10		47	**9**	**7**	**1**	**3**	10	10
24	**4**	**9**	**6**	**1**	10	10		48	**9**	**8**	**1**	**2**	10	10

If for instance, the 2,4,8,6 Number Set is used in a Complementary Frame, then only 2 out of the 6 Number Sequences namely the 2,4,8,6 and 2,6,8,4 Number Sequences will return four (4) Complementary Frame Positions each. The 2,4,6,8 and 2,8,6,4 and 2,6,4,8 and 2,8,4,6 Number Sequences cannot form a Complementary Frame Position. The 2,4,8,6 and 2,6,8,4 Number Sequences represents the Forward Clockwise and the Backwards Anticlockwise Direction Movement Positions respectively. These are shown in the tables below. Note that the numbers on the BR column are complements of the numbers on the TL Column. Likewise, the numbers on the BL column are complements of the numbers on the TR Column.

	TL	TR	BR	BL
C1	2	4	6	8
A1	2	8	6	4
1				
2				
3				
4				
5				
6				
7				
8				

	TL	TR	BR	BL
C2	2	4	8	6
A2	2	6	8	4
1	**2**	**4**	**8**	**6**
2	**6**	**2**	**4**	**8**
3	**8**	**6**	**2**	**4**
4	**4**	**8**	**6**	**2**
5	**2**	**6**	**8**	**4**
6	**4**	**2**	**6**	**8**
7	**8**	**4**	**2**	**6**
8	**6**	**8**	**4**	**2**

	TL	TR	BR	BL
C3	2	6	4	8
A3	2	8	4	6
1				
2				
3				
4				
5				
6				
7				
8				

Grouping of the X-Frame Subcomponent Number Sequences

As already mentioned, a pair of Number Sequence in every Magic Square Grid Frame has eight (8) positions, four (4) Clockwise positions and four (4) Anticlockwise Positions irrespective of the order or size of the Magic Square Grid, however, some pair of Number Sequences do not necessarily form Eight (8) unique Frame Positions after rotation or reflection. Such pair of Number Sequences can only have Six (6), four (4), or one (1) unique Frame Position, depending on the Complementary and Number Repeat Specification of the Magic Square Grid Number Range. For instance, the 2,2,8,8 and 5,5,5,5 Number Sequences can only form four (4) unique Frame Positions and one (1)unique Frame Position respectively.

1

2		2
8		8

2

8		2
8		2

3

8		8
2		2

4

2		8
2		8

5

2		8
8		2

6

8		2
2		8

7

8

The 2,2,8,8 number set can only form one unique Position.

5		5
5		5

The 5,5,5,5 number set can only form one unique Position.
Therefore, the X-Frame Sub components are divided into Groups based on the number of unique positions that can be formed with the number selections. The 5,5,5,5 selection is

known as the **1-position X-Frame Subcomponent** since it is only one and does not belong to any group.

X-FRAME SUBCOMPONENT GROUPING			
GROUP NUMBER	**NUMBER REPEAT**	**MAXIMUM NUMBER OF NON-COMPLEMENTARY POSITIONS OF NUMBER SET**	**MAXIMUM NUMBER OF COMPLEMENTARY POSITIONS OF NUMBER SEQUENCE**
1	4 different. Numbers, Used 1 times each.	24	8
2	3 different. Numbers, 1 Repeated 2 times.	12	0
3	2 Different Numbers repeated 2 times Each	6	0
4	2 different. Numbers, 1 Repeated 3 times.	4	0
5	1 Number Repeated 4 times.	1	1

X-Frame Subcomponent Identifier Table

In order to properly identify the different Number Sequences from different Number Sets according to the Number of positions that can be possibly formed from the different groups, it is necessary to insert another table beside the X-frame Subcomponent table. This table is known as the X-Frame Subcomponent Identifier table. An example of the X-Frame Subcomponent identifier Table is shown below.

X-Frame Subcomponent Identifier Table						X-Frame Subcomponent Table			
X-Comp S/N	GROUP	NUMBER SET S/N	NUMBER SEQUENCE S/N	POSITION		TL	TR	BR	BL
13	3	8		1	-	2	4	8	6
15	3	8		5	-	2	6	8	4
29	3	8		8	-	4	2	6	8
35	3	8		4	-	4	8	6	2
47	3	8		2	-	6	2	4	8
53	3	8		6	-	6	8	4	2
67	3	8		7	-	8	4	2	6
69	3	8		3	-	8	6	2	4

The Vertical and Serial X-Frame Subcomponents Table Rearrangement

The X-Frame Subcomponents table can be rearranged in such that all the possible Eighty-One Positions are lined-up in a particular order. The positions can either be lined up according to the X-Frame Subcomponent Selections or they can as well be lined up serially in ascending order. As shown in the two tables below. These rearrangement will be of importance to us in the subsequent chapters while filling-in the Plus-Component numbers in the different frames of Magic Grid. Either of the tables can be chosen at a time.

Table 1									**Table 2**							
VERTICAL X-FRAME SUBCOMPONENT RE-ARRANGEMENT									**SERIAL X-FRAME SUBCOMPONENT RE-ARRANGEMENT**							
S/N	**Group**	**Selection**	**Position**	**T L**	**T R**	**B R**	**B L**	**Possible Unique Grid Positions**	S/N	**Group**	**Selection**	**Position**	**T L**	**T R**	**B R**	**B L**
1	1	1	1	**1**	**1**	**9**	**9**	4	1	1	1	1	**1**	**1**	**9**	**9**
2	1	1	2	**9**	**1**	**1**	**9**		2	2	2	1	**1**	**2**	**9**	**8**
3	1	1	3	**9**	**9**	**1**	**1**		3	2	3	1	**1**	**3**	**9**	**7**
4	1	1	4	**1**	**9**	**9**	**1**		4	2	4	1	**1**	**4**	**9**	**6**
5	2	2	1	**1**	**2**	**9**	**8**	8	5	1	5	1	**1**	**5**	**9**	**5**
6	2	2	2	**8**	**1**	**2**	**9**		6	2	4	5	**1**	**6**	**9**	**4**
7	2	2	3	**9**	**8**	**1**	**2**		7	2	3	5	**1**	**7**	**9**	**3**
8	2	2	4	**2**	**9**	**8**	**1**		8	2	2	5	**1**	**8**	**9**	**2**

9	2	2	5	**1**	**8**	**9**	**2**	
10	2	2	6	**8**	**9**	**2**	**1**	
11	2	2	7	**9**	**2**	**1**	**8**	
12	2	2	8	**2**	**1**	**8**	**9**	
13	2	3	1	**1**	**3**	**9**	**7**	8
14	2	3	2	**7**	**1**	**3**	**9**	
15	2	3	3	**9**	**7**	**1**	**3**	
16	2	3	4	**3**	**9**	**7**	**1**	
17	2	3	5	**1**	**7**	**9**	**3**	
18	2	3	6	**7**	**9**	**3**	**1**	
19	2	3	7	**9**	**3**	**1**	**7**	
20	2	3	8	**3**	**1**	**7**	**9**	
21	2	4	1	**1**	**4**	**9**	**6**	8
22	2	4	2	**6**	**1**	**4**	**9**	
23	2	4	3	**9**	**6**	**1**	**4**	
24	2	4	4	**4**	**9**	**6**	**1**	
25	2	4	5	**1**	**6**	**9**	**4**	
26	2	4	6	**6**	**9**	**4**	**1**	
27	2	4	7	**9**	**4**	**1**	**6**	
28	2	4	8	**4**	**1**	**6**	**9**	
29	1	5	1	**1**	**5**	**9**	**5**	4
30	1	5	2	**5**	**1**	**5**	**9**	
31	1	5	3	**9**	**5**	**1**	**5**	
32	1	5	4	**5**	**9**	**5**	**1**	
33	1	6	1	**2**	**2**	**8**	**8**	4
34	1	6	2	**8**	**2**	**2**	**8**	
35	1	6	3	**8**	**8**	**2**	**2**	
36	1	6	4	**2**	**8**	**8**	**2**	
37	2	7	1	**2**	**3**	**8**	**7**	8
38	2	7	2	**7**	**2**	**3**	**8**	

9	1	1	4	**1**	**9**	**9**	**1**
10	2	2	8	**2**	**1**	**8**	**9**
11	1	6	1	**2**	**2**	**8**	**8**
12	2	7	1	**2**	**3**	**8**	**7**
13	2	8	1	**2**	**4**	**8**	**6**
14	1	9	1	**2**	**5**	**8**	**5**
15	2	8	5	**2**	**6**	**8**	**4**
16	2	7	5	**2**	**7**	**8**	**3**
17	1	6	4	**2**	**8**	**8**	**2**
18	2	2	4	**2**	**9**	**8**	**1**
19	2	3	8	**3**	**1**	**7**	**9**
20	2	7	8	**3**	**2**	**7**	**8**
21	1	10	1	**3**	**3**	**7**	**7**
22	2	11	1	**3**	**4**	**7**	**6**
23	1	12	1	**3**	**5**	**7**	**5**
24	2	11	5	**3**	**6**	**7**	**4**
25	1	10	4	**3**	**7**	**7**	**3**
26	2	7	4	**3**	**8**	**7**	**2**
27	2	3	4	**3**	**9**	**7**	**1**
28	2	4	8	**4**	**1**	**6**	**9**
29	2	8	8	**4**	**2**	**6**	**8**
30	2	11	8	**4**	**3**	**6**	**7**
31	1	13	1	**4**	**4**	**6**	**6**
32	1	14	1	**4**	**5**	**6**	**5**
33	1	13	4	**4**	**6**	**6**	**4**
34	2	11	4	**4**	**7**	**6**	**3**
35	2	8	4	**4**	**8**	**6**	**2**
36	2	4	4	**4**	**9**	**6**	**1**
37	1	5	2	**5**	**1**	**5**	**9**
38	1	9	2	**5**	**2**	**5**	**8**

39	2	7	3	**8**	**7**	**2**	**3**	
40	2	7	4	**3**	**8**	**7**	**2**	
41	2	7	5	**2**	**7**	**8**	**3**	
42	2	7	6	**7**	**8**	**3**	**2**	
43	2	7	7	**8**	**3**	**2**	**7**	
44	2	7	8	**3**	**2**	**7**	**8**	
45	2	8	1	**2**	**4**	**8**	**6**	
46	2	8	2	**6**	**2**	**4**	**8**	
47	2	8	3	**8**	**6**	**2**	**4**	
48	2	8	4	**4**	**8**	**6**	**2**	8
49	2	8	5	**2**	**6**	**8**	**4**	
50	2	8	6	**6**	**8**	**4**	**2**	
51	2	8	7	**8**	**4**	**2**	**6**	
52	2	8	8	**4**	**2**	**6**	**8**	
53	1	9	1	**2**	**5**	**8**	**5**	
54	1	9	2	**5**	**2**	**5**	**8**	4
55	1	9	3	**8**	**5**	**2**	**5**	
56	1	9	4	**5**	**8**	**5**	**2**	
57	1	10	1	**3**	**3**	**7**	**7**	
58	1	10	2	**7**	**3**	**3**	**7**	4
59	1	10	3	**7**	**7**	**3**	**3**	
60	1	10	4	**3**	**7**	**7**	**3**	
61	2	11	1	**3**	**4**	**7**	**6**	
62	2	11	2	**6**	**3**	**4**	**7**	
63	2	11	3	**7**	**6**	**3**	**4**	
64	2	11	4	**4**	**7**	**6**	**3**	8
65	2	11	5	**3**	**6**	**7**	**4**	
66	2	11	6	**6**	**7**	**4**	**3**	
67	2	11	7	**7**	**4**	**3**	**6**	
68	2	11	8	**4**	**3**	**6**	**7**	

39	1	12	2	**5**	**3**	**5**	**7**
40	1	14	2	**5**	**4**	**5**	**6**
41	0	15	1	**5**	**5**	**5**	**5**
42	1	14	4	**5**	**6**	**5**	**4**
43	1	12	4	**5**	**7**	**5**	**3**
44	1	9	4	**5**	**8**	**5**	**2**
45	1	5	4	**5**	**9**	**5**	**1**
46	2	4	2	**6**	**1**	**4**	**9**
47	2	8	2	**6**	**2**	**4**	**8**
48	2	11	2	**6**	**3**	**4**	**7**
49	1	13	2	**6**	**4**	**4**	**6**
50	1	14	3	**6**	**5**	**4**	**5**
51	1	13	3	**6**	**6**	**4**	**4**
52	2	11	6	**6**	**7**	**4**	**3**
53	2	8	6	**6**	**8**	**4**	**2**
54	2	4	6	**6**	**9**	**4**	**1**
55	2	3	2	**7**	**1**	**3**	**9**
56	2	7	2	**7**	**2**	**3**	**8**
57	1	10	2	**7**	**3**	**3**	**7**
58	2	11	7	**7**	**4**	**3**	**6**
59	1	12	3	**7**	**5**	**3**	**5**
60	2	11	3	**7**	**6**	**3**	**4**
61	1	10	3	**7**	**7**	**3**	**3**
62	2	7	6	**7**	**8**	**3**	**2**
63	2	3	6	**7**	**9**	**3**	**1**
64	2	2	2	**8**	**1**	**2**	**9**
65	1	6	2	**8**	**2**	**2**	**8**
66	2	7	7	**8**	**3**	**2**	**7**
67	2	8	7	**8**	**4**	**2**	**6**
68	1	9	3	**8**	**5**	**2**	**5**

69	1	12	1	**3**	**5**	**7**	**5**	4	69	2	8	3	**8**	**6**	**2**	**4**
70	1	12	2	**5**	**3**	**5**	**7**		70	2	7	3	**8**	**7**	**2**	**3**
71	1	12	3	**7**	**5**	**3**	**5**		71	1	6	3	**8**	**8**	**2**	**2**
72	1	12	4	**5**	**7**	**5**	**3**		72	2	2	6	**8**	**9**	**2**	**1**
73	1	13	1	**4**	**4**	**6**	**6**	4	73	1	1	2	**9**	**1**	**1**	**9**
74	1	13	2	**6**	**4**	**4**	**6**		74	2	2	7	**9**	**2**	**1**	**8**
75	1	13	3	**6**	**6**	**4**	**4**		75	2	3	7	**9**	**3**	**1**	**7**
76	1	13	4	**4**	**6**	**6**	**4**		76	2	4	7	**9**	**4**	**1**	**6**
77	1	14	1	**4**	**5**	**6**	**5**	4	77	1	5	3	**9**	**5**	**1**	**5**
78	1	14	2	**5**	**4**	**5**	**6**		78	2	4	3	**9**	**6**	**1**	**4**
79	1	14	3	**6**	**5**	**4**	**5**		79	2	3	3	**9**	**7**	**1**	**3**
80	1	14	4	**5**	**6**	**5**	**4**		80	2	2	3	**9**	**8**	**1**	**2**
81	0	15	1	**5**	**5**	**5**	**5**	1	81	1	1	3	**9**	**9**	**1**	**1**

The Plus-Frame Subcomponent

In the previous section, we dealt with the X Frame Subcomponent of a Magic Square Grid, in this section we will look at the Plus Subcomponent.

The Plus-Frame Subcomponent are the numbers in the squares between each pair of the four diagonal-end squares (X-Frame Subcomponents) in an opposite manner of a particular Magic Square Grid Frame.

The number of squares between each pair of the four diagonal-end squares a Frame is equal to the subtraction of "2" from the Magic Square Order "n" of that particular Frame, and such this portion of the Magic Square Grid Frame is referred to as n-2.

The members of the Plus-Frame subcomponent are so numerous that most of the variations of Magic Squares are derived from it.

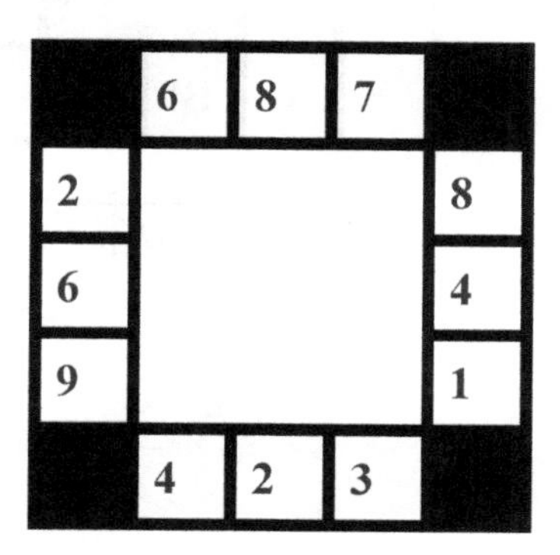

The Plus-Frame Sub component are made up of two distinct parts, namely; The Vertical Plus-Frame Subcomponent (Vn-2) and The Horizontal Plus-Frame Subcomponent (Hn-2).

The Vertical Plus-Frame Subcomponent (Vn-2) is also made up of the Top Vertical Plus-Frame subcomponent (TVn-2) and the Down Vertical Plus-Frame Subcomponent (DVn-2) while the Horizontal Plus-Frame Subcomponent (Hn-2) is made up of the Right Horizontal Plus-Frame Subcomponent (RHn-2) and the Horizontal Plus-Frame Subcomponent (LHn-2)

a. <u>The Vertical Plus-Frame Subcomponent (Vn-2)</u>

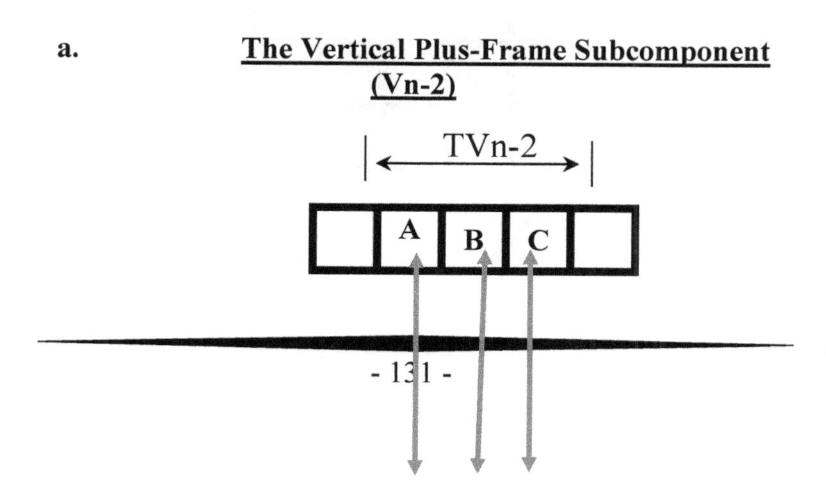

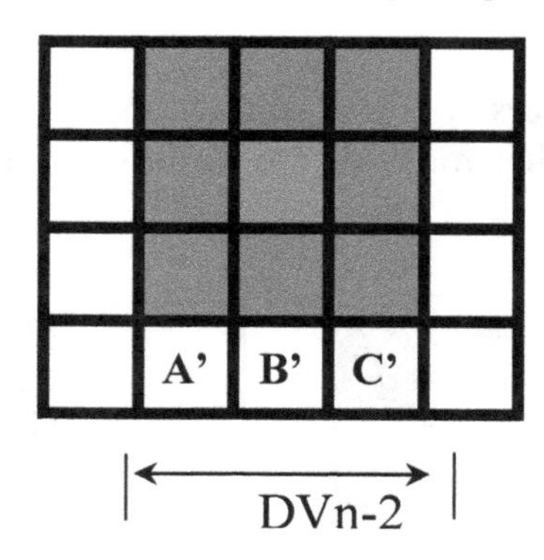

b. **The Horizontal Plus-Frame Subcomponent (Hn-2)**

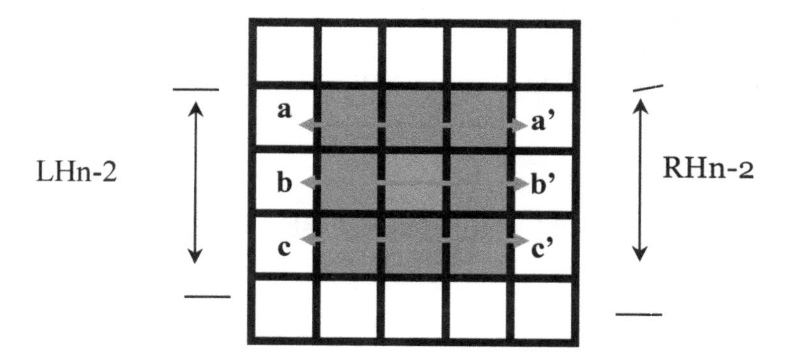

Methods of Generating the Plus-Frame Subcomponent Numbers.

The Plus component members can be filled in two ways namely;

1) The Permutations of Restricted Integer Partitions Method.
2) The Restricted Integer Compositions Method.

*The Permutations of Restricted Integer Partitions Method entails generating the **Permutations of the Restricted Integer Partitions** of the number difference between the Magic Sum and X-Frame Subcomponent members on each side of the Frame.*

While

*The Restricted Integer Compositions Method entails Generating the **Restricted Integer Compositions** of the number difference between the Magic Sum and the X-Frame Subcomponent members on each side of the Frame.*

Integer Partitions and Compositions

In number theory and Combinatorics, a **partition** of a positive integer *n*, also called an **integer partition**, is a way of writing *n* as a sum of positive integers.

Two sums that differ only in the order of their summands are considered the same partition but different compositions.

For example, 4 can be partitioned in five distinct ways
4, 3 + 1, 2 + 2, 2 + 1 + 1, 1 + 1 + 1 + 1.

The order-dependent **composition** 1 + 3 is the same partition as 3 + 1, while 1 + 2 + 1 and 1 + 1 + 2 are the same partition as 2 + 1 + 1.

i.e. they are the same partition but different compositions.

A summand in a partition is also called a **part**. The notation $\lambda \vdash n$ means that λ is a partition of n.

Restricted Integer Partitions and Compositions

A restricted partition is a partition in which the parts are constrained in some way, while a restricted Composition of a number is a composition in which the members are constrained in the same way as the Partitions.

There are several types of restricted Partitions and Compositions. For example, we could count partitions that contain only odd numbers. Among the 22 partitions of the number 8, there are 6 that contain only *odd parts*:

- 7 + 1
- 5 + 3
- 5 + 1 + 1 + 1
- 3 + 3 + 1 + 1
- 3 + 1 + 1 + 1 + 1 + 1
- 1 + 1 + 1 + 1 + 1 + 1 + 1 + 1

Alternatively, we could count partitions in which no number occurs more than once. If we count the partitions of 8 with *distinct parts*, we also obtain 6:

- 8
- 7 + 1
- 6 + 2
- 5 + 3

- 5 + 2 + 1
- 4 + 3 + 1

Application of Permutations, Restricted Integer Partitions/ Compositions in Generating the Plus Component of Magic Square Grids.

A special kind of Restricted Integer Partition and Composition is required in generating the Plus Component of Magic Square Grids.

The type of Restricted Partition or composition required is the type in which only the Partitions or Compositions that contain only a given number of Parts or summands are selected among all the possible partitions or compositions.

Mathematically expressed;

*The Restricted Integer Partition or Composition of a Magic Square Grid Frame has to do with the partitions of **n** in which the number count of Parts are equal, and corresponds to the number of squares between each side of the four diagonal-end squares of the particular Frame in the Magic Square Grid.*

In Magic Square Grids Partitioning, We are concerned with the various permuted partitions or compositions that can fill the gap between the two diagonal end squares on each side of a Magic Square Grid Frame (X Frame Subcomponents). Each permuted partition or Composition forms a new unique Magic Square Grid Frame.

As mentioned at the beginning of this section, all the possible Plus Subcomponent members of a Particular Frame can be worked out either by **Permuting** the already generated **Restricted Integer Partitions** of a number **or by** generating the **Restricted Integer Compositions** of the number, both of which will give the same result.

However, the first method entails generating the different Integer Partitions of the number, after which a permutation of each of the Partitions are derived While the second method involves generating the Integer Compositions of the number straight. Therefore, we can conclude that:

"The total number of Integer Compositions of a number is the same as the Permutation of the Integer Partitions of that same number."

By the end of this section, we will see how generating the Permutations of the Integer partitions of a number is same as generating the Compositions of that number.

Note that the essence of generating the Restricted Integer Compositions of the number or Permutation of the generated Restricted Integer Partitions of the number is because a change in the order of the parts forms a new unique Magic Square Frame. I.e. Every composition of a Partition forms a new Magic Square Grid Variation.

You will find out in Complementary Magic Square Grids Frames that the numbers in the squares that are in-between any two diagonal-end squares when paired in an opposite manner have a special characteristic in that they can be permuted to form a new Magic Square.

Number of Magic Square Grid Plus Component Permutations.

From the diagrams below, A & A', B & B', C & C' and a & a', b & b', c & c', are permutable.

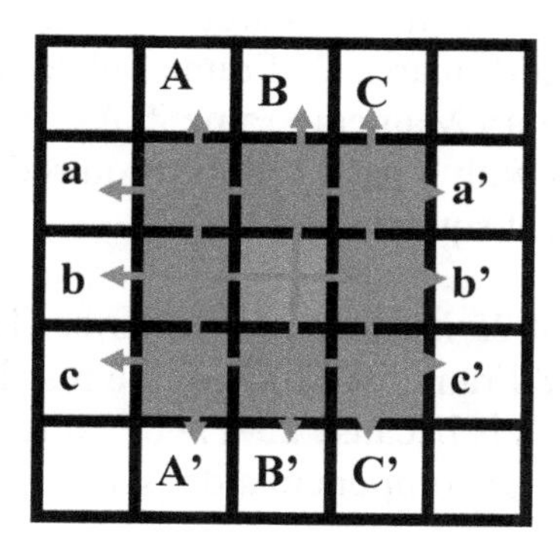

In Magic Square theory, number of Squares per Column or Row is known as Order (***n***).

In Magic Square Grids, Number of Permutable Squares (***n***-2) is equivalent to number of elements (n). Likewise, Number of Chosen Permutable Squares (k***n***-2) is equivalent to number of Chosen Elements (k).

Note the difference between the "n" in Permutation and the "***n***" which stands for "Order" in Magic Square theory.

Therefore:
Permutation = P(n-2), kn-2
= (n-2)! / ((n-2)- kn-2)!

If applied to Magic Square Grids, Number of Permutable Squares (***n***-2) is equal to number of chosen Permutable Squares (kn-2).

This is because n=k (as in Permutation)

Therefore:
Permutation = Factorial
= P ($\boldsymbol{n}$-2), k n-2 = ($\boldsymbol{n}$-2)!

For Instance, there are three squares between each pair of the four Diagonal-End-Squares (X-Frame Subcomponent) of an Odd Complementary Magical Frame 2. As such, only the Integer Partitions with 3 parts will be required for filling this Plus subcomponent squares.

If a Magic Sum of 25 is chosen for instance, Then, the 8,9 side of the 1,2,9,8 Frame will require the restricted integer partitions of 8, this is because 25-17=8. Note that the sum of the 8,9 X-Frame Subcomponent side is 17.
i.e. 25 - (8+9) = 8

Therefore we need to generate the Permutation of a Three (3) Part Restricted Integer Partitions of 8, or the Compositions of 8, in order to know all the possibilities of filling the empty squares.

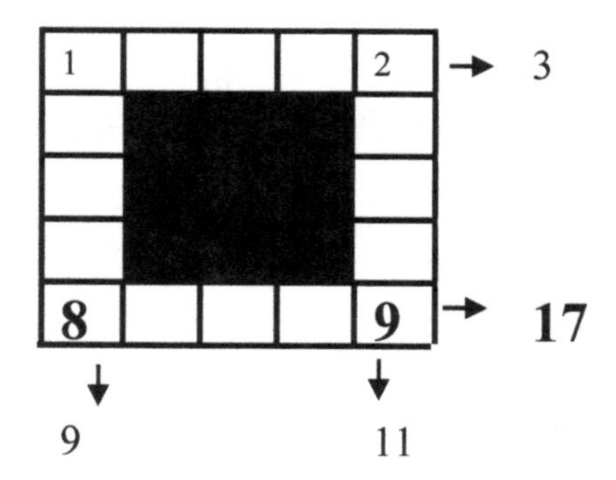

Integer Partitions of Number 8

8
7 + 1
6+2
6 + 1 + 1
5+3
5 + 2 + 1
5 + 1 + 1 + 1
4 + 4
4+3+1
4+2+2
4 + 2 + 1 + 1
4+1+1+1+1
3+3+2
3+3+1+1
3+2+2+1
3+2+1+1+1
3 + 1 + 1 + 1 + 1 + 1
2+2+2+2
2+2+2+1+1
2+2+1+1+1+1
2+1+1+1+1+1+1
1+1+1+1+1+1+1+1

Note that the remaining partitions other than the ones listed here are compositions, which are permutations of the listed partitions and therefore considered the same partition as the listed.

Three (3) Part Restricted Integer Partitions of 8

From the listed partitions of 8 above, 5 partitions have only 3 parts that sums up to 17 and required to complete the 8,9 side of the frame shown above.

6 + 1 + 1

5 + 2 + 1
4 + 3 + 1
4 + 2 + 2
3 + 3 + 2

Using the 6 + 1 + 1 Partition of 8 as an example;

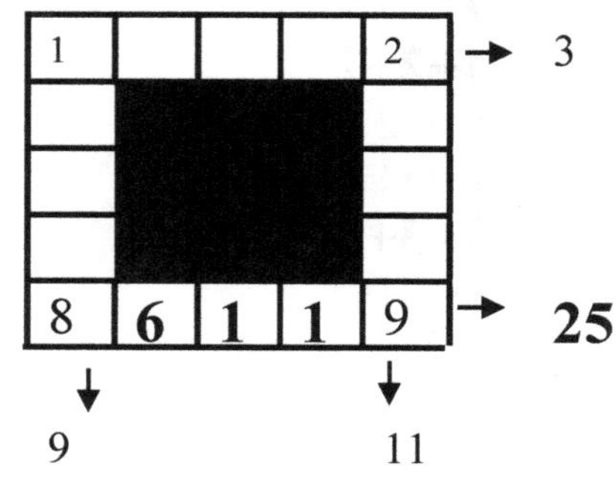

If a Permutation of each of the five 3-Part Restricted Integer Partitions are generated, then, we will have 21 Permutations, thus,

For the **6+1+1** Partition, we have

1) 6 + 1 + 1
2) 1 + 6 + 1
3) 1 + 1 + 6

For the **5+2 +1** Partition, we have
4) 5 + 2 + 1
5) 5 + 1 + 2
6) 2 + 5 + 1
7) 2 + 1 + 5
8) 1 + 5 + 2

9) 1 + 2 + 5

For the **4+3+1** Partition, we have
10) 4 + 3 + 1
11) 4 + 1 + 3
12) 3 + 4 + 1
13) 3 + 1 + 4
14) 1 + 4 + 3
15) 1 + 3 + 4

For the **4+2+2** Partition, we have
16) 4 + 2 + 2
17) 2 + 4 + 2
18) 2 + 2 + 4

For the **3+3+2** Partition, we have
19) 3 + 3 + 2
20) 3 + 2 + 3
21) 2 + 3 + 3

Three (3) Part Restricted Integer Composition of 8

On the other hand, if a 3-Part Restricted Integer Composition are generated, we will still have a total of 21 Compositions, thus;

The complete list of the three part integer compositions of 8 is shown below.

1)	6	+	1	+	1
2)	5	+	1	+	2
3)	5	+	2	+	1
4)	4	+	1	+	3
5)	4	+	2	+	2

6) 4 + 3 + 1
7) 3 + 1 + 4
8) 3 + 2 + 3
9) 3 + 3 + 2
10) 3 + 4 + 1
11) 2 + 1 + 5
12) 2 + 2 + 4
13) 2 + 3 + 3
14) 2 + 4 + 2
15) 2 + 5 + 1
16) 1 + 1 + 6
17) 1 + 2 + 5
18) 1 + 3 + 4
19) 1 + 4 + 3
20) 1 + 5 + 2
21) 1 + 6 + 1

TABLE OF POSSIBLE INTEGER COMPOSITIONS OF THE FIRST FIVE (5) ODD NUMBERED FRAMES OF MAGIC SQUARE GRIDS. i.e. Frame 1,2,3,4 and 5

S/N	Frame 1		Frame 2		Frame 3		Frame 4		Frame 5	
	Integer Compositions Sum	Number of 1-Part Compositions	Integer CompositionsSum	Number of 3-Part Compositions	Integer Compositions Sum	Number of 5-Part Compositions	Integer Compositions Sum	Number of 7-Part Compositions	Integer Compositions Sum	Number of 9-Part Compositions
1	1	1	7	15	17	1,645	27	144,186	37	11,854,197
2	2	1	8	21	18	2,030	28	167,826	38	13,352,247

3	3	1	9	28	19	2,430	29	191,226	39	14,795,247
4	4	1	10	36	20	2,826	30	213,402	40	16,131,483
5	6	1	11	45	21	3,195	31	233,331	41	17,309,772
6	7	1	12	52	22	3,510	32	250,026	42	18,282,597
7	8	1	13	57	23	3,750	33	262,626	43	19,009,197
8	9	1	14	60	24	3,900	34	270,466	44	19,458,297
9			15	61	25	3,951	35	273,127	45	19,610,233
10			16	60	26	3,900	36	270,466	46	19,458,297
11			17	57	27	3,750	37	262,626	47	19,009,197
12			18	52	28	3,510	38	250,026	48	18,282,597
13			19	45	29	3,195	39	233,331	49	17,309,772
14			20	36	30	2,826	40	213,402	50	16,131,483
15			21	28	31	2,430	41	191,226	51	14,795,247
16			22	21	32	2,030	42	167,826	52	13,352,247
17			23	15	33	1,645	43	144,186	53	11,854,197

TABLE OF POSSIBLE INTEGER COMPOSITIONS OF THE SECOND FIVE (5) FRAMES. i.e. Frame 6, 7, 8, 9, 10

S/N	Frame 6		Frame 7		Frame 8		Frame 9		Frame 10	
	Integer Componsitions Sum	Number of 11-Part Compositions	Integer Componsitions Sum	Number of 13-Part Compositions	Integer Componsitions Sum	Number of 15-Part Compositions	Integer Componsitions Sum	Number of 17-Part Compositions	Integer Componsitions Sum	Number of 19-Part Compositions (Approximated)
1	47	951,247,935	57	75,545,988,921	67	5,972,330,545,875	77	471,265,969,202,244	87	37,167,582,884,496,100
2	48	1,049,164,875	58	82,109,870,817	68	6,421,397,664,375	78	502,503,337,343,376	88	39,370,520,104,891,800
3	49	1,141,816,005	59	88,242,672,921	69	6,837,025,296,795	79	531,203,380,941,888	89	41,382,697,885,324,900

4	50	1,226,323,593	60	93,776,561,853	70	7,209,054,894,735	80	556,732,609,361,352	90	43,163,635,350,443,400
5	51	1,299,917,322	61	98,553,025,974	71	7,528,026,744,360	81	578,507,503,698,801	91	44,676,366,452,174,900
6	52	1,360,080,117	62	102,430,202,277	72	7,785,569,661,915	82	596,016,192,262,905	92	45,888,690,047,160,100
7	53	1,404,686,547	63	105,289,679,469	73	7,974,757,067,895	83	608,838,011,673,849	93	46,774,286,922,035,600
8	54	1,432,123,407	64	107,042,334,165	74	8,090,408,810,115	84	616,659,932,481,273	94	47,313,650,711,667,300
9	55	1,441,383,219	65	107,632,809,909	75	8,129,320,828,911	85	619,288,973,447,049	95	47,494,787,636,620,700
10	56	1,432,123,407	66	107,042,334,165	76	8,090,408,810,115	86	616,659,932,481,273	96	47,313,650,711,667,300
11	57	1,404,686,547	67	105,289,679,469	77	7,974,757,067,895	87	608,838,011,673,849	97	46,774,286,922,035,600
12	58	1,360,080,117	68	102,430,202,277	78	7,785,569,661,915	88	596,016,192,262,905	98	45,888,690,047,160,100
13	59	1,299,917,322	69	98,553,025,974	79	7,528,026,744,360	89	578,507,503,698,801	99	44,676,366,452,174,900
14	60	1,226,323,593	70	93,776,561,853	80	7,209,054,894,735	90	556,732,609,361,352	100	43,163,635,350,443,400
15	61	1,141,816,005	71	88,242,672,921	81	6,837,025,296,795	91	531,203,380,941,888	101	41,382,697,885,324,900
16	62	1,049,164,875	72	82,109,870,817	82	6,421,397,664,375	92	502,503,337,343,376	102	39,370,520,104,891,800
17	63	951,247,935	73	75,545,988,921	83	5,972,330,545,875	93	471,265,969,202,244	103	37,167,582,884,496,100

Mathematical and Statistical Tools for generating Partitions, Compositions, Permutations and Combinations.

The "R" Programming Language

Ever heard of the "R" Programming Language? R is a programming language and software environment for statistical computing and graphics. You can read more about the R Programming Language at the CRAN website; ***http://cran.r-project.org***or at;***http://www.r-project.org/***
or at; Wikipedia;
http://en.wikipedia.org/wiki/R_(programming_language)

There are many ways of generating Integer Partitions and one good way is to use the Partitions Package of the R language and the partitions package. In this session, we are going to look at how to make a "Partitions Calculator" with the R Programming language, using the Partitions Package. Not only will we be able to know the number of Partitions or Compositions of a number but we will also be able to generate a list of the Integer Partitions or Compositions. Download and install the R language which is a free software available at **http://cran.r-project.org** also download and install the Partitions Package.Open the program and type the following codes;

require(partitions)
parts(5)

the following result will be given;
[1,] 5 4 3 3 2 2 1
[2,] 0 1 2 1 2 1 1
[3,] 0 0 0 1 1 1 1
[4,] 0 0 0 0 0 1 1
[5,] 0 0 0 0 0 0 1
Where each column is a partition of 5.

More information on how to use the R Language in generating Partitions can be found at the following URL;

http://cran.r project.org/web/packages/partitions/partitions.pdf
then

How to generate the restricted Parts

Now we will try to generate the Restricted Parts of an Integer that contains only single digits. i.e. restricted Partitions of an Integer with 9 as the largest Parts and 0 are excluded.

Generally, this is expressed by the function;

"restrictedparts(n, m,include.zero=FALSE, decreasing=TRUE)"

but a recursion, can be used in generating a more clean result for a
partition of **n** with **m** positive parts where the largest part is no more
than**k** can be made from a partition of **n-x** with **m-1** parts where the largest part is no more than **x** (for any **x** with **n/m<=x<=k**) by adding a new largest part of **x**. However, you may need to generate a large number of smaller partitions.

As an alternative in R, if the numbers are not too big, you could for example find partitions of 14 with 2 positive parts where the largest part is no more than 9 with the following codes;

```
require(partitions)
pbig<- as.matrix(restrictedparts(14, 2, include.zero=FALSE))
psmall<- pbig[, pbig[1,] <= 9]
```

sopbig is

```
     [,1] [,2] [,3] [,4] [,5] [,6] [,7]
[1,]  13   12   11   10    9    8    7
[2,]   1    2    3    4    5    6    7
```

While psmall is

```
     [,1] [,2] [,3]
[1,]    9    8    7
[2,]    5    6    7
```

This will be inefficient and memory intensive when **n** gets too big: so for partitions of 100 into 12 positive parts, pbig will have 5994463 partitions while psmall only has 22 where the largest part is no more than 9.

Use of the matrixStats package

Download and install the matrixStats package, then type the following;

```
require(partitions)
require(matrixStats)
pbig<- as.matrix(compositions(20, m=3, include.zero=FALSE))
psmall<- pbig[, colMaxs(pbig) <= 9]
psmall
```

You should get 36 compositions.

The Sectional Component

Two or more distinct Magic Square Grids can be combined to form another bigger Magic Square Grid, each of the Magic Square Grid that forms part of the bigger grid is known as a section.

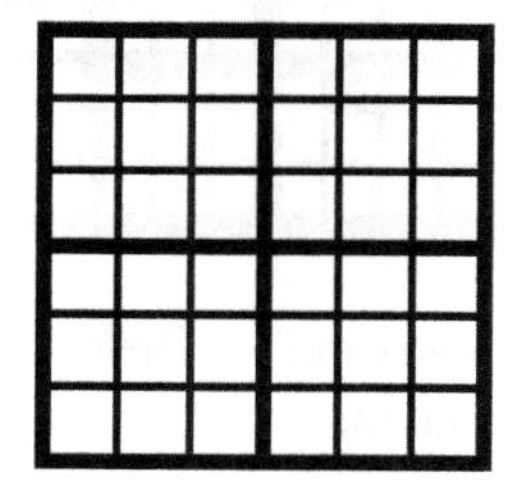

Order-6 4-Multi-Sectional Order-3 Magic Square Grid.

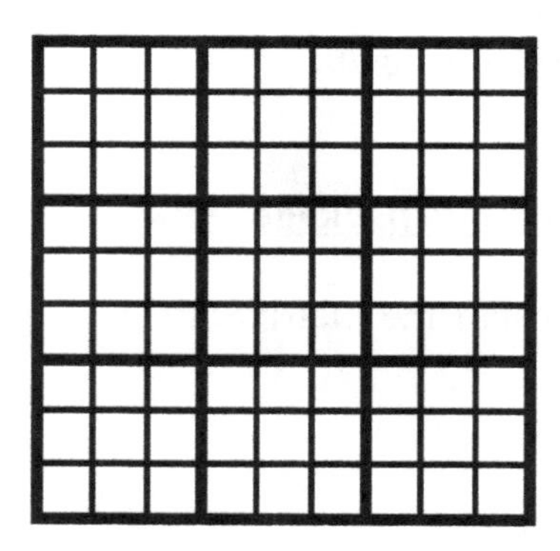

Order-9 9-Multi-Sectional Order-3 Magic Square Grid.

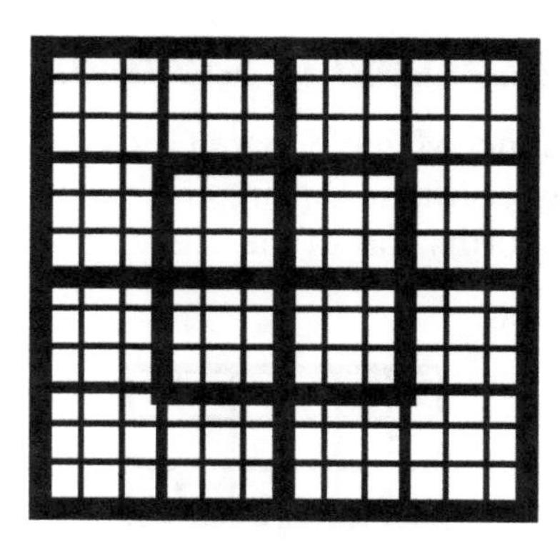

Order-12 16-Multi-Sectional Order-3 5-Multi-Sectional Order-6 Magic Square Grid.

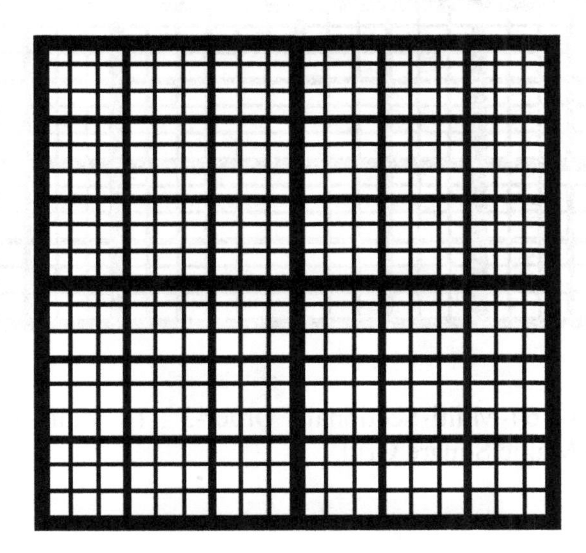

Order-18 36-Multi-Sectional Order-3 4-Multi-Sectional Order-9 Magic Square Grid.

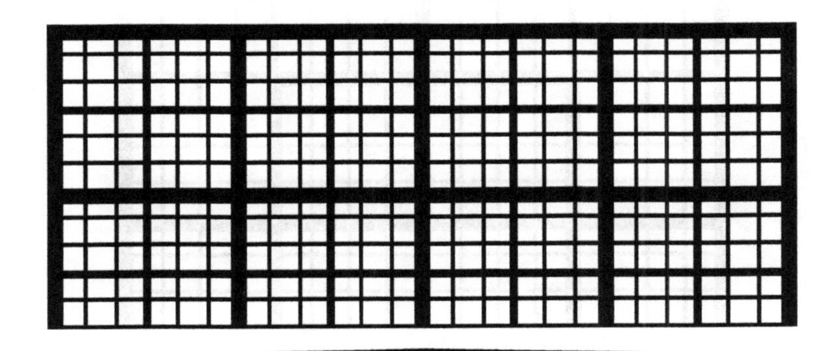

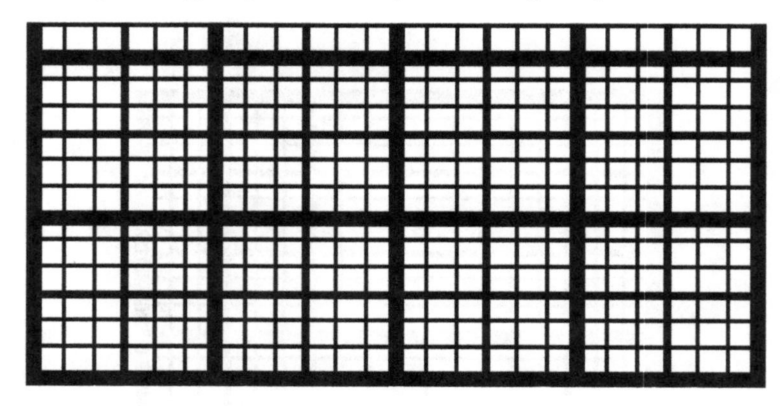

Order-24 64-Multi-Sectional Order-3 16-Multi-Sectional Order-6 Magic Square Grid.

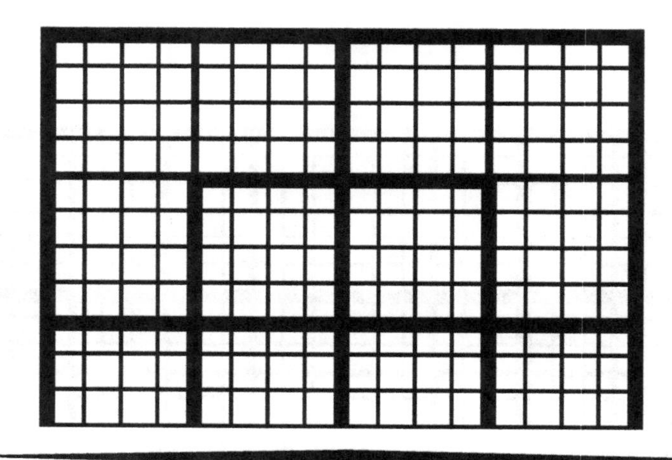

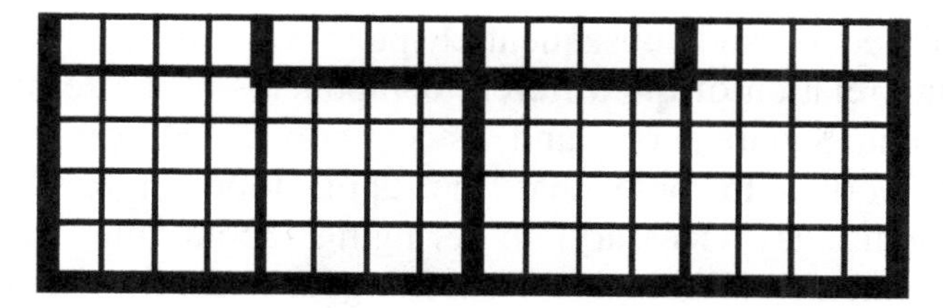

Order-16; 16-Multi-Sectional,Order-4; 5-Multi-Sectional, Order-8; Magic Square Grid.

Inter-Components Relationship in the Formation of Afamatrix Magic Square Grid Variations.

So far, in this Chapter, we have learnt about the four (4) Minor Components of Magic Square Grids namely; the Cells, the Rows, the Columns, and the Diagonals which is made up of the Forward-Slash Diagonal and Backward-Slash Diagonal. We have also learnt about the Two (2) Major Components of Magic Square Grids namely the Sectional Component, and the Frame Component made up of Plus-Frame Subcomponent and X-Frame Subcomponent. On the other hand, we have also seen how numbers can be generated for the different Frame subcomponents.

In this section and subsequent chapters we will be looking at the interrelation of the different components in the formation of Magic Square Grids and its Variations. However, it is most worthy to note that the application of Deductive Reasoning is necessary in bringing about the number arrangements. Deductive Reasoning in the sense that our conclusions are logically based on the agreement of multiple grounds that are generally assumed to be true. The first topic on Inter Cell relationship will enable us have a better understanding of the relationship that exist between the numbers in the cells, the importance is that it will help us find a quicker way of Constructing Magic Square Grids.

Inter Cell Relationship

In the Frame 1 of a perfect Odd order Afamatrix Magic Square Grid which is the outermost frame of the 3 x 3 grid or level 1 Afamatrix, the arrangement of the numbers in the frame together with the number 5 added at the center square is typical of a Lo shu Square or its number transposition. There is only one (1) perfect form, but there are eight (8) possible variations. These variations are either a Rotation or Reflection of the perfect form. When starting the arrangement of the frame 1 of an Afamatrix Magic Square Grid, it is important to first consider the flow and interrelationship of the X component Groups and the number selections. If a complimentary Frame is being constructed, then these set of numbers are arranged in such a way as to create balance in the frame. Bearing in mind about the compliment pair of numbers, i.e. the pair of numbers that adds up to 10.

2	**9**	**4**	→	**15**
7	**5**	**3**	→	**15**
6	**1**	**8**	→	**15**

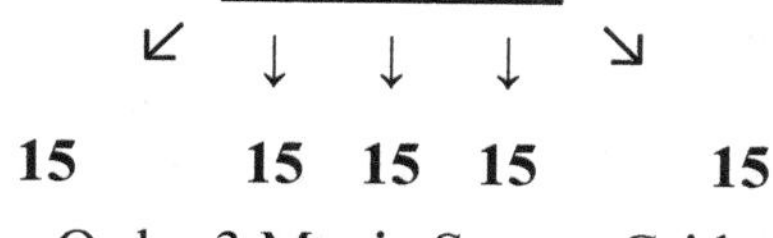

Order 3 Magic Square Grid

For Order 4, the following relationships exist;
A+B = C+D in all the cases.

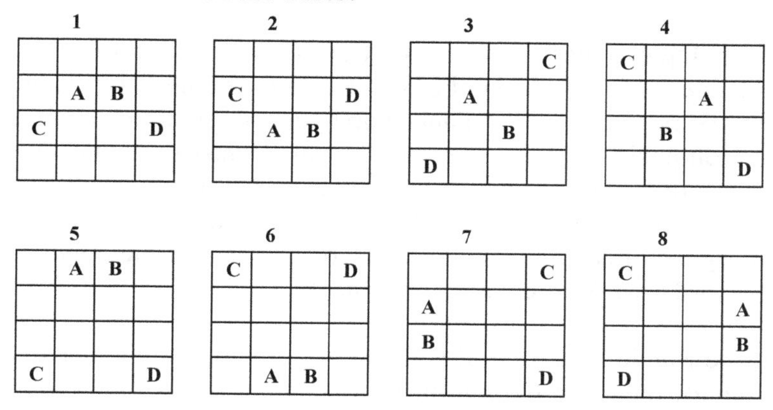

Remember that the X component Group numbers are filled inside the frame in an “X” form inside the 4 diagonal ends squares also known as the Cardinal points. **I.e.** the Top Leftmost Square, the Top Rightmost Square, the Bottom Leftmost Square and the Bottom Rightmost Square.

Using the Afam’s Method of Magic Square Grid Construction, we will find out that a new Magic Square Grid is formed whenever there is a change in any **Frame Position** of a Perfect Magic Square Grid. i.e Rotation or Reflection of a Frame. (Refer to the section below on Frame Positioning for a better understanding of Magic Square Grid Frame Positions).

On the other hand, a new Magic Square Grid is formed whenever there is a change in Composition in any of the Horizontal or Vertical Plus Subcomponent part of any of the Frames that make up the Magic Square Grid. These changes in composition is dependent upon the various X-Frame Subcomponent Number Sequences on any of the Magic Square Grid Frames that make up the Magic Square Grid.

For Magic Square Grids with Complementary Frames, the Change in Composition of any of the Left or Right Horizontal Plus Subcomponent is balanced up by a corresponding change in Composition of the opposite Left or Right Horizontal Plus Subcomponent. Likewise the Change in Composition of any of the Top or Down Vertical Plus Subcomponent is balanced up by a corresponding change in Composition of the opposite Top or Down Vertical Subcomponent.

If the Magic Square Grid is made up of Non-Complementary Frames, then, the Change in Composition of any of the Left or Right Horizontal Plus Subcomponent is balanced up by a corresponding change in Composition of any of the Left or Right Horizontal Plus Subcomponent of any Frame that make up the Magic Square Grid, but not necessarily the opposite Left or Right Horizontal Plus Subcomponent of the same Frame.

Likewise, the Change in Composition of any of the Top or Down Vertical Plus Subcomponent is balanced up by a corresponding change in Composition of the Top or Down Vertical Subcomponent of any Frame that make up the Magic Square Grid, but not necessarily the opposite Top or Down Vertical Plus Subcomponent of the same Frame

CHAPTER 8

Part 2 - Afam's Method of Constructing Magic Square Grids

Application of Tables in the Formation of Magic Square Grids

The Afam's Method of forming Magic Square Grids is more of the construction of Magic Square Grid Tables. Once we understand how to construct all the required tables, then we are over Ninety Percent (90%) done. The remaining less than Ten Percent (10%) job is all about transferring numbers from the tables to the actual Magic Square Grid. Introduction to the basics of these Tables has been handled at the Beginning of

Chapter 6. In this Chapter more tables will be used in showing the application of Magic Square Grid Tables in the formation of Magic Square Grids.

The first step in constructing Magic Square Grids using the Afam's method is to make a **Comprehensive Magic Square Grid Parameters Table** in which the required Properties of the needed Magic Square Grid are listed. Unlike the Basic Magic Square Grid Parameters Table, the Comprehensive Magic Square Grid Parameters Table contains the complete details of the Magic Square Grid to be constructed. As you can see from the Table below, more parameters covering the Plus Component aspect of the Frame has been added. These includes the Frame Number , Plus Component Composition Parts, the Plus Component Table Range which includes the Highest Plus Component Number and the Highest Plus Component Number expected in the range, etc.

Shown below is the comprehensive Magic Square Grid Properties Table.

S/N	***Comprehensive Magic Square Grid Parameters Table***				
	PARAMETERS	**SUB-PARAMETERS**	**VALUES**		
1	GENERAL	Grid Type	Semi-Perfect		
		Order / Level	3 /1.5		
2	POSSIBLE GRID MAGICSUM RANGE	Smallest Possible			
		Largest Possible			
		Chosen			
3	SECTIONS	Type			
		Interchangeability			
		Possible Sections			
		Section Order			
4	THE FRAMES	Frame Number	0.0	1.0	2.0
		Frame Type	Comp.	Comp.	Non-

					Comp
5	NUMBER RANGE	Smallest			
		Largest			
		Specific Numbers Only			
		Exception Number			
		Exception Num. Qty			
		Number Repeat			
		Max Repeat Qty			
6	POSSIBLE MAGIC SUM RANGE OF DIFFERENT LEVELS	Smallest	20	30	40
		Largest	24	34	44
		Chosen	20	30	40
7	X-FRAME SUBCOMPONENT TABLE RANGE	Type			
		Lowest			
		Highest			
8	PLUS-FRAME SUBCOMPONENT INTEGER COMPOSITION TABLE RANGE	Lowest Limit Table			
		Highest Limit Table			
		Lowest Composition			
		Highest Composition			

According the specifications of the Magic Square Grid Parameters Table, a number of Tables will need to be constructed in other to find all the variations of the specified Magic Square Grid. Listed below are the tables that will be required be required.

1) Single Center-Square Number (for Odd Order), or Centre Square-Table (for Even Order).
2) X-Frame Sub Components Table
3) Integer Compositions Selection Tables or Plus-Frame Subcomponent selection Table
4) Frame Computation Table
5) Magic Square Grid Variations Table.

Each Table listed is also accompanied with Identifier Table and Checking Tables whenever necessary. The making of

Identifier Tables and Checking Tables has been treated in the previous chapter. In order for us to be able to construct the Frame Computation Table and Magic Square Grid Variations Table listed in number 4 and 5 above, there is need for us to have an ample knowledge of **Cross Joined Tables**. A Cross Join Table is formed by the combination of rows from two or more tables. The resultant table of a Cross Join is the product of the sets of records from the two of more joined tables. For example, the third table below is a Cross Join Table of the first two tables.

Table 1

1	A	B
2	C	D
3	E	F

Table 2

1	U	V
2	W	X
3	Y	Z

Table 3

1	A	B	U	V
2	A	B	W	X
3	A	B	Y	Z
4	C	D	U	V
5	C	D	W	X
6	C	D	Y	Z
7	E	F	U	V
8	E	F	W	X
9	E	F	Y	Z

The details on the making of the tables are as follows;

1) Single Center-Square Number, or Centre Square-Table

In Afam's method, the Center Square Number or Center Square Zero (0) comprises mostly of a single digit inside one square being the first odd order in Magic square Grids, while Centre Square 0.5 comprises of 4 numbers inside four squares being the first even order of Magic Square Grids.

The Single Center-Square Number is for Odd Order Magic Square Grids while the Centre Square-Table is for Even Order Magic Square Grids. The Centre Square-Table could be the same as the X-Frame Sub Components Table depending on the specified parameters of the Magic Square Grid. As already mentioned in the beginning, the case of a single digit inside a single square is considered to be trivial, and for Magic Square Grids of Even Order, Order 2 is also trivial. I.e. it is impossible to form a Magic Square Grid of Order 2. Order 2 is referred to as Center Square 0.5 in Afam's Method. The smallest non-trivial case is the 3 x 3 Magic Square Grid.

In order to have the first non-trivial case (Order 3), Centre Square Zero is always joined together with Frame 1, while Centre Square 0.5 is always joined with Frame 1.5 in order to have the first non-trivial case of even order.

1

Odd Center Square"0"

1	**2**
3	**4**

Even Center Square "0"

2) X-Frame Subcomponent Table

The details of the making of the X-Frame Subcomponent Table containing all the possible X-Frame Subcomponent Number Sequences has been explained in the previous Chapter.
Either the Vertical Table or the Serial Vertical table can be used.

X-Frame Subcomponent Table

S/N	Group	Selection	Position	TL	TR	BR	BL
1	1	1	1	1	1	9	9
2	2	2	1	1	2	9	8
3	2	3	1	1	3	9	7
4	2	4	1	1	4	9	6
5	1	5	1	1	5	9	5
77	1	5	3	9	5	1	5
78	2	4	3	9	6	1	4
79	2	3	3	9	7	1	3
80	2	2	3	9	8	1	2
81	1	1	3	9	9	1	1

3) <u>Integer Compositions Selection Tables(Plus-Frame Subcomponent Selection Tables).</u>

These are Series of Tables containing the Integer Compositions of the Plus Frame Sub Component. It is made up of the Lower-Limit Table and Upper-Limit Table together with other in-between Tables. The needed Integer Compositions Selection Tables are determined from the specified Number Range and Magic Sum of the Magic Square Grid.

In a case where number repeat is allowed,
Lower Limit = Magic Sum – (2 * Largest Range Number)
While
Upper Limit = Magic Sum – (2 * Smallest Range Number)

But in a case where number repeat is not allowed, then
Lower Limit = Magic Sum – (Largest Range Number + Next Largest Range Number)
While
Upper Limit = Magic Sum – (Smallest Range Number + next Smallest Range Number)

The Integer Compositions Selection Table has to be drawn because there are several possible number selection that can fit into the empty squares between the X Frame Subcomponent. For instance, if we are to construct the Integer Compositions Selections Tables needed to find all the possible variations of an Order 4 Magic Square Grid with a Magic Sum of 20 when the specified number range is 1-9 repeatable, then;

9			**9**
1			**1**

Lower Limit = Magic Sum – (2 * Largest Range Number)
= 20 – (2*9)
= 20 – (18) = 2

Therefore, the Lower Limit Table will be made up of Integer Composition of 2.

Likewise;
Upper Limit = Magic Sum – (2 * Smallest Range Number)
= 20 – (2*1)
= 20 – (2) = 18

Therefore, the Upper-Limit Table will be made up of Integer Composition of 18.

The Complete range of Tables will be between 2 – 18 as shown below.

2-Part Compositions of Frame 1.5

Magic Sum = 20, Number Range = 1-9 Repeatable

A	B	C	D	E	F
I	I.C.	I.C.	I.C.	I.C.	I.
.	S	S	S	S	C.
C	=3	=4	=5	=6	S

	A	B
1	1	1

	A	B
1	1	2
2	2	1

	A	B
1	1	3
2	2	2
3	3	1

	A	B
1	1	4
2	2	3
3	3	2
4	4	1

	A	B
1	1	5
2	2	4
3	3	3
4	4	2
5	5	1

	A	B
1	1	6
2	2	5
3	3	4
4	4	3
5	5	2
6	6	1

G
I.C.
S
=8

	A	B
1	1	7
2	2	6
3	3	5
4	4	4
5	5	3
6	6	2
7	7	1

H
I.C.
S
=9

	A	B
1	1	8
2	2	7
3	3	6
4	4	5
5	5	4
6	6	3
7	7	2
8	8	1

I
I.C.
S
=10

	A	B
1	1	9
2	2	8
3	3	7
4	4	6
5	5	5
6	6	4
7	7	3
8	8	2
9	9	1

J
I.C.
S
=11

	A	B
1	2	9
2	3	8
3	4	7
4	5	6
5	6	5
6	7	4
7	8	3
8	9	2

K
I.C.
S
=12

	A	B
1	3	9
2	4	8
3	5	7
4	6	6
5	7	5
6	8	4
7	9	3

L
I
.
C

	A	B
1	4	9
2	5	8

M
I.C.
S
=14

	A	B
1	5	9
2	6	8

N
I.C.
S
=15

	A	B
1	6	9
2	7	8

O
I.C.
S
=16

	A	B
1	7	9
2	8	8

P
I.C.
S
=17

	A	B
1	8	9
2	9	8

Q
I.
C.
S

	A	B
1	9	9

3	**6**	**7**
4	**7**	**6**
5	**8**	**5**
6	**9**	**4**

3	**7**	**7**
4	**8**	**6**
5	**9**	**5**

3	**8**	**7**
4	**9**	**6**

3	**9**	**7**

Likewise, if we are to find the Integer Compositions Selections Tables needed to construct all the possible variations of an Order 4 Magic Square Grid with a Magic Sum of 20 when the specified number range is 1-9 **non-repeatable**, then;

9			**8**
2			**1**

Lower Limit = Magic Sum – (Largest Range Number + Next Largest Range Number)
= 20 – (9+8)
= 20 – (17) = 3

Therefore, the Lower Limit Table will be made up of Integer Composition of 3.

Also;
Upper Limit = Magic Sum – (Smallest Range Number + Nest Smallest Range Number)
= 20 – (1 + 2)
= 20 – (3) = 17

Therefore, the Upper-Limit Table will be made up of Integer Composition of 17.

The Complete range of Tables will then be between 3 – 17. (All the tables shown above with the exception of Table A and Table B)

4) Frame Computation Table

A **Frame Computation Table** containing all the possible Variations of a Frame is made by listing the Upper and Lower Horizontal Plus Subcomponent Compositions in relation with the Left and Right Vertical Plus subcomponent Compositions, together with each of the X-Frame Components with which the Compositions are made.

For Non-Complementary Magic Square Grid Frames, the Integer Compositions of every side of the Frames are filled independently and cross-joined with each other, but for Complementary Magic Square Grid Frames, the Integer Compositions of the Topmost Row and the Leftmost Column are first written down after which the corresponding opposite vacant squares are written down. The vertical rows are then cross joined with the horizontal rows. Since a kind of balance exists within every Magic Square Grid, you will discover that after filling-in the compliment numbers in the opposite vacant squares in the bottommost row and the rightmost column, you must have as well selected the required pair of numbers necessary for completing the Bottommost row and the Rightmost column.
Mathematically,
Total Possible Frame Variations=Number of X-Frame Number Sequences * Number of Horizontal Plus-Frame Subcomponent Compositions * Number of Vertical Plus-Frame Subcomponent Compositions.

Example:

FRAME COMPUTATION TABLE									
Frame Number =?									
Computation S/N	**X Component**				**Plus-Component Compositions (Permutations)**				**Type (Frame)**
					Vertical		**Horizontal**		
	Group	**Selection**	**Position**	**Number**	**Up Side**	**Down Side**	**Left Side**	**Right Side**	
					X-Comp Sum =5 + 5 =10 Integer Compositions Sum =25-10=15	X-Comp Sum =5 + 5 =10 Integer Compositions Sum =25-10=15	X-Comp Sum =5 + 5 =10 Integer Compositions Sum =25-10=15	X-Comp Sum =5 + 5 =10 Integer Compositions Sum =25-10=15	
1	1	1	1	**2-4-8-6**	**9 + 4 + 2**	**1 + 6 + 8**	**7 + 4 + 4**	**3 + 6 + 6**	Perfect
2	1	1	1	**2-4-8-6**	**9 + 4 + 2**	**1 + 6 + 8**	**7 + 6 + 2**	**3 + 4 + 8**	Perfect
3	1	1	1	**2-4-8-6**	**9 + 4 + 2**	**1 + 6 + 8**	**7 + 5 + 3**	**3 + 5 + 7**	Imperfect
4	1	1	1	**2-4-8-6**	**9 + 4 + 2**	**1 + 6 + 8**	**7 + 4 + 4**	**3 + 6 + 6**	

FRAME 1 (Showing X-Frame Subcomponents with only Hyper-Afamatrix Magic Square Grid formation Capabilities and X-Frame Subcomponents with both Afamatrix and Hyper-Afamatrix Magic Square Grid formation Capabilities)				
X-Componen t Identifier	**X- Component**	**Plus-Component**		**CLAS S**
		VERTICA L	**HORIZNT L**	

X-Comp S/N	Group	Selection	Position	T L	T R	B R	B L	T.S.	D.S.	L.S.	R.S.	
				1	2	1'	2'	A	A'	A	a'	
1	2	1	1	1	1	9	9	13	-3	5	5	H-Afamatrix
2	3	2	1	1	2	9	8	12	-2	6	4	H-Afamatrix
3	3	3	1	1	3	9	7	11	-1	7	3	H-Afamatrix
4	3	4	1	1	4	9	6	10	0	8	2	H-Afamatrix
5	2	5	1	1	5	9	5	9	1	9	1	Afamatrix
6	3	4	5	1	6	9	4	8	2	10	0	H-Afamatrix
7	3	3	5	1	7	9	3	7	3	11	-1	H-Afamatrix
8	3	2	5	1	8	9	2	6	4	12	-2	H-Afamatrix
9	2	1	4	1	9	9	1	5	5	13	-3	H-Afamatrix
10	3	2	8	2	1	8	9	12	-2	4	6	H-Afamatrix
11	2	6	1	2	2	8	8	11	-1	5	5	H-Afamatrix
12	3	7	1	2	3	8	7	10	0	6	4	H-Afamatrix
13	3	8	1	2	4	8	6	9	1	7	3	Afamatrix
14	2	9	1	2	5	8	5	8	2	8	2	Afamatrix
15	3	8	5	2	6	8	4	7	3	9	1	Afamatrix
16	3	7	5	2	7	8	3	6	4	10	0	H-Afamatrix
17	2	6	4	2	8	8	2	5	5	11	-1	H-Afamatrix
18	3	2	4	2	9	8	1	4	6	12	-2	H-Afamatrix
19	3	3	8	3	1	7	9	11	-1	3	7	H-Afamatrix
20	3	7	8	3	2	7	8	10	0	4	6	H-Afamatrix
21	2	10	1	3	3	7	7	9	1	5	5	Afamatrix
22	3	11	1	3	4	7	6	8	2	6	4	Afamatrix
23	2	12	1	3	5	7	5	7	3	7	3	Afamatrix
24	3	11	5	3	6	7	4	6	4	8	2	Afamatrix
25	2	10	4	3	7	7	3	5	5	9	1	Afamatrix
26	3	7	4	3	8	7	2	4	6	10	0	H-Afamatrix

27	3	3	4	3	9	7	1	3	7	11	-1	H-Afamatrix
28	3	4	8	4	1	6	9	10	0	2	8	H-Afamatrix
29	3	8	8	4	2	6	8	9	1	3	7	Afamatrix
30	3	11	8	4	3	6	7	8	2	4	6	Afamatrix
31	2	13	1	4	4	6	6	7	3	5	5	Afamatrix
32	2	14	1	4	5	6	5	6	4	6	4	Afamatrix
33	2	13	4	4	6	6	4	5	5	7	3	Afamatrix
34	3	11	4	4	7	6	3	4	6	8	2	Afamatrix
35	3	8	4	4	8	6	2	3	7	9	1	Afamatrix
36	3	4	4	4	9	6	1	2	8	10	0	H-Afamatrix
37	2	5	2	5	1	5	9	9	1	1	9	Afamatrix
38	2	9	2	5	2	5	8	8	2	2	8	Afamatrix
39	2	12	2	5	3	5	7	7	3	3	7	Afamatrix
40	2	14	2	5	4	5	6	6	4	4	6	Afamatrix
41	1	15	1	5	5	5	5	5	5	5	5	Afamatrix
42	2	14	4	5	6	5	4	4	6	6	4	Afamatrix
43	2	12	4	5	7	5	3	3	7	7	3	Afamatrix
44	2	9	4	5	8	5	2	2	8	8	2	Afamatrix
45	2	5	4	5	9	5	1	1	9	9	1	Afamatrix
46	3	4	2	6	1	4	9	8	2	0	10	H-Afamatrix
47	3	8	2	6	2	4	8	7	3	1	9	Afamatrix
48	3	11	2	6	3	4	7	6	4	2	8	Afamatrix
49	2	13	2	6	4	4	6	5	5	3	7	Afamatrix
50	2	14	3	6	5	4	5	4	6	4	6	Afamatrix
51	2	13	3	6	6	4	4	3	7	5	5	Afamatrix
52	3	11	6	6	7	4	3	2	8	6	4	Afamatrix
53	3	8	6	6	8	4	2	1	9	7	3	Afamatrix
54	3	4	6	6	9	4	1	0	10	8	2	H-Afamatrix
55	3	3	2	7	1	3	9	7	3	-1	11	H-Afamatrix
56	3	7	2	7	2	3	8	6	4	0	10	H-Afamatrix

57	2	10	2	7	3	3	7	5	5	1	9	Afamatrix
58	3	11	7	7	4	3	6	4	6	2	8	Afamatrix
59	2	12	3	7	5	3	5	3	7	3	7	Afamatrix
60	3	11	3	7	6	3	4	2	8	4	6	Afamatrix
61	2	10	3	7	7	3	3	1	9	5	5	Afamatrix
62	3	7	6	7	8	3	2	0	10	6	4	H-Afamatrix
63	3	3	6	7	9	3	1	-1	11	7	3	H-Afamatrix
64	3	2	2	8	1	2	9	6	4	-2	12	H-Afamatrix
65	2	6	2	8	2	2	8	5	5	-1	11	H-Afamatrix
66	3	7	7	8	3	2	7	4	6	0	10	H-Afamatrix
67	3	8	7	8	4	2	6	3	7	1	9	Afamatrix
68	2	9	3	8	5	2	5	2	8	2	8	Afamatrix
69	3	8	3	8	6	2	4	1	9	3	7	Afamatrix
70	3	7	3	8	7	2	3	0	10	4	6	H-Afamatrix
71	2	6	3	8	8	2	2	-1	11	5	5	H-Afamatrix
72	3	2	6	8	9	2	1	-2	12	6	4	H-Afamatrix
73	2	1	2	9	1	1	9	5	5	-3	13	H-Afamatrix
74	3	2	7	9	2	1	8	4	6	-2	12	H-Afamatrix
75	3	3	7	9	3	1	7	3	7	-1	11	H-Afamatrix
76	3	4	7	9	4	1	6	2	8	0	10	H-Afamatrix
77	2	5	3	9	5	1	5	1	9	1	9	Afamatrix
78	3	4	3	9	6	1	4	0	10	2	8	H-Afamatrix
79	3	3	3	9	7	1	3	-1	11	3	7	H-Afamatrix
80	3	2	3	9	8	1	2	-2	12	4	6	H-Afamatrix
81	2	1	3	9	9	1	1	-3	13	5	5	H-Afamatrix

FRAME 1 PERFECT AFAMATRIX MAGIC SQUARE GRID TABLE
(Showing only the Perfect Possibilities)

ect E1 me 1	X-	X-	Plus-	CL

		Component Identifier				Component				Component				AS S
										VERTICAL		HORIZONTAL		
		X-Comp S/N	Group	Selectn	Positn	T L	T R	B R	B L	T. S.	D .S.	L. S.	R. S.	
						1	2	1'	2'	A	A'	A	a'	
1	2	13	3	8	1	2	4	8	6	9	1	7	3	Perfect
2	4	15	3	8	5	2	6	8	4	7	3	9	1	Perfect
3	10	29	3	8	8	4	2	6	8	9	1	3	7	Perfect
4	16	35	3	8	4	4	8	6	2	3	7	9	1	Perfect
5	26	47	3	8	2	6	2	4	8	7	3	1	9	Perfect
6	32	53	3	8	6	6	8	4	2	1	9	7	3	Perfect
7	38	67	3	8	7	8	4	2	6	3	7	1	9	Perfect
8	40	69	3	8	3	8	6	2	4	1	9	3	7	Perfect

FRAME 1 AFAMATRIX MAGIC SQUARE GRID TABLE (Showing the Perfect and Imperfect Possibilities)					
Frame 1 S/N	X- Component Identifier	X-Component	Plus-Component		CLASS
			VERTICAL	HORIZONTAL	

	X-Comp S/N	Group	Selectn	Positn	TL	TR	BR	BL	T.S.	D.S.	L.S.	R.S.	
					1	2	1'	2'	A	A'	A	a'	
1	5	2	5	1	1	5	9	5	9	1	9	1	Imperfect
2	13	3	8	1	2	4	8	6	9	1	7	3	Perfect
3	14	2	9	1	2	5	8	5	8	2	8	2	Imperfect
4	15	3	8	5	2	6	8	4	7	3	9	1	Perfect
5	21	2	10	1	3	3	7	7	9	1	5	5	Imperfect
6	22	3	11	1	3	4	7	6	8	2	6	4	Imperfect
7	23	2	12	1	3	5	7	5	7	3	7	3	Imperfect
8	24	3	11	5	3	6	7	4	6	4	8	2	Imperfect
9	25	2	10	4	3	7	7	3	5	5	9	1	Imperfect
10	29	3	8	8	4	2	6	8	9	1	3	7	Perfect
11	30	3	11	8	4	3	6	7	8	2	4	6	Imperfect
12	31	2	13	1	4	4	6	6	7	3	5	5	Imperfect
13	32	2	14	1	4	5	6	5	6	4	6	4	Imperfect
14	33	2	13	4	4	6	6	4	5	5	7	3	Imperfect
15	34	3	11	4	4	7	6	3	4	6	8	2	Imperfect
16	35	3	8	4	4	8	6	2	3	7	9	1	Perfect
17	37	2	5	2	5	1	5	9	9	1	1	9	Imperfect
18	38	2	9	2	5	2	5	8	8	2	2	8	Imperfect
19	39	2	12	2	5	3	5	7	7	3	3	7	Imperfect
20	40	2	14	2	5	4	5	6	6	4	4	6	Imperfect
21	41	1	15	1	5	5	5	5	5	5	5	5	Imperfect
22	42	2	14	4	5	6	5	4	4	6	6	4	Imperfect

23	43	2	12	4	5	7	5	3	3	7	7	3	Imperfect
24	44	2	9	4	5	8	5	2	2	8	8	2	Imperfect
25	45	2	5	4	5	9	5	1	1	9	9	1	Imperfect
26	47	3	8	2	6	2	4	8	7	3	1	9	Perfect
27	48	3	11	2	6	3	4	7	6	4	2	8	Imperfect
28	49	2	13	2	6	4	4	6	5	5	3	7	Imperfect
29	50	2	14	3	6	5	4	5	4	6	4	6	Imperfect
30	51	2	13	3	6	6	4	4	3	7	5	5	Imperfect
31	52	3	11	6	6	7	4	3	2	8	6	4	Imperfect
32	53	3	8	6	6	8	4	2	1	9	7	3	Perfect
33	57	2	10	2	7	3	3	7	5	5	1	9	Imperfect
34	58	3	11	7	7	4	3	6	4	6	2	8	Imperfect
35	59	2	12	3	7	5	3	5	3	7	3	7	Imperfect
36	60	3	11	3	7	6	3	4	2	8	4	6	Imperfect
37	61	2	10	3	7	7	3	3	1	9	5	5	Imperfect
38	67	3	8	7	8	4	2	6	3	7	1	9	Perfect
39	68	2	9	3	8	5	2	5	2	8	2	8	Imperfect
40	69	3	8	3	8	6	2	4	1	9	3	7	Perfect
41	77	2	5	3	9	5	1	5	1	9	1	9	Imperfect

FRAME 2 AFAMATRIX MAGIC SQUARE GRID TABLE (Showing the first 100 out of the 185,997 Possibilities) Note the First 100 possibilities are all Imperfect

F2 S/N	X- Component Identifier			X- Component				Plus-Component				CLASS
								VERTICAL		HORIZONTAL		
	X- Comp S/N	GROUP	SELECTION	POSITION	TL	TR	BR	BL	TOP-SIDE	DOWN-SIDE	LEFT-SIDE	RIGHT-SIDE

					1	1'	2	2'	A	B	C	A'	B'	C'	a	b	c	a'	b'	c'	
1	1	2	1	1	1	1	9	9	9	9	5	1	1	5	9	5	1	1	5	9	Imp
2	1	2	1	1	1	1	9	9	9	9	5	1	1	5	9	4	2	1	6	8	Imp
3	1	2	1	1	1	1	9	9	9	9	5	1	1	5	9	3	3	1	7	7	Imp
4	1	2	1	1	1	1	9	9	9	9	5	1	1	5	9	2	4	1	8	6	Imp
5	1	2	1	1	1	1	9	9	9	9	5	1	1	5	9	1	5	1	9	5	Imp
6	1	2	1	1	1	1	9	9	9	9	5	1	1	5	8	6	1	2	4	9	Imp
7	1	2	1	1	1	1	9	9	9	9	5	1	1	5	8	5	2	2	5	8	Imp
8	1	2	1	1	1	1	9	9	9	9	5	1	1	5	8	4	3	2	6	7	Imp
9	1	2	1	1	1	1	9	9	9	9	5	1	1	5	8	3	4	2	7	6	Imp
10	1	2	1	1	1	1	9	9	9	9	5	1	1	5	8	2	5	2	8	5	Imp
11	1	2	1	1	1	1	9	9	9	9	5	1	1	5	8	1	6	2	9	4	Imp
12	1	2	1	1	1	1	9	9	9	9	5	1	1	5	7	7	1	3	3	9	Imp
13	1	2	1	1	1	1	9	9	9	9	5	1	1	5	7	6	2	3	4	8	Imp
14	1	2	1	1	1	1	9	9	9	9	5	1	1	5	7	5	3	3	5	7	Imp
15	1	2	1	1	1	1	9	9	9	9	5	1	1	5	7	4	4	3	6	6	Imp
16	1	2	1	1	1	1	9	9	9	9	5	1	1	5	7	3	5	3	7	5	Imp
17	1	2	1	1	1	1	9	9	9	9	5	1	1	5	7	2	6	3	8	4	Imp
18	1	2	1	1	1	1	9	9	9	9	5	1	1	5	7	1	7	3	9	3	Imp
19	1	2	1	1	1	1	9	9	9	9	5	1	1	5	6	8	1	4	2	9	Imp
20	1	2	1	1	1	1	9	9	9	9	5	1	1	5	6	7	2	4	3	8	Imp
21	1	2	1	1	1	1	9	9	9	9	5	1	1	5	6	6	3	4	4	7	Imp
22	1	2	1	1	1	1	9	9	9	9	5	1	1	5	6	5	4	4	5	6	Imp
23	1	2	1	1	1	1	9	9	9	9	5	1	1	5	6	4	5	4	6	5	Imp
24	1	2	1	1	1	1	9	9	9	9	5	1	1	5	6	3	6	4	7	4	Imp
25	1	2	1	1	1	1	9	9	9	9	5	1	1	5	6	2	7	4	8	3	Imp
26	1	2	1	1	1	1	9	9	9	9	5	1	1	5	6	1	8	4	9	2	Imp

27	1	2	1	1	1	1	9	9	9	9	5	1	1	5	5	9	1	5	1	9	Imp
28	1	2	1	1	1	1	9	9	9	9	5	1	1	5	5	8	2	5	2	8	Imp
29	1	2	1	1	1	1	9	9	9	9	5	1	1	5	5	7	3	5	3	7	Imp
30	1	2	1	1	1	1	9	9	9	9	5	1	1	5	5	6	4	5	4	6	Imp
31	1	2	1	1	1	1	9	9	9	9	5	1	1	5	5	5	5	5	5	5	Imp
32	1	2	1	1	1	1	9	9	9	9	5	1	1	5	5	4	6	5	6	4	Imp
33	1	2	1	1	1	1	9	9	9	9	5	1	1	5	5	3	7	5	7	3	Imp
34	1	2	1	1	1	1	9	9	9	9	5	1	1	5	5	2	8	5	8	2	Imp
35	1	2	1	1	1	1	9	9	9	9	5	1	1	5	5	1	9	5	9	1	Imp
36	1	2	1	1	1	1	9	9	9	9	5	1	1	5	4	9	2	6	1	8	Imp
37	1	2	1	1	1	1	9	9	9	9	5	1	1	5	4	8	3	6	2	7	Imp
38	1	2	1	1	1	1	9	9	9	9	5	1	1	5	4	7	4	6	3	6	Imp
39	1	2	1	1	1	1	9	9	9	9	5	1	1	5	4	6	5	6	4	5	Imp
40	1	2	1	1	1	1	9	9	9	9	5	1	1	5	4	5	6	6	5	4	Imp
41	1	2	1	1	1	1	9	9	9	9	5	1	1	5	4	4	7	6	6	3	Imp
42	1	2	1	1	1	1	9	9	9	9	5	1	1	5	4	3	8	6	7	2	Imp
43	1	2	1	1	1	1	9	9	9	9	5	1	1	5	4	2	9	6	8	1	Imp
44	1	2	1	1	1	1	9	9	9	9	5	1	1	5	3	9	3	7	1	7	Imp
45	1	2	1	1	1	1	9	9	9	9	5	1	1	5	3	8	4	7	2	6	Imp
46	1	2	1	1	1	1	9	9	9	9	5	1	1	5	3	7	5	7	3	5	Imp
47	1	2	1	1	1	1	9	9	9	9	5	1	1	5	3	6	6	7	4	4	Imp
48	1	2	1	1	1	1	9	9	9	9	5	1	1	5	3	5	7	7	5	3	Imp
49	1	2	1	1	1	1	9	9	9	9	5	1	1	5	3	4	8	7	6	2	Imp
50	1	2	1	1	1	1	9	9	9	9	5	1	1	5	3	3	9	7	7	1	Imp
51	1	2	1	1	1	1	9	9	9	9	5	1	1	5	2	9	4	8	1	6	Imp
52	1	2	1	1	1	1	9	9	9	9	5	1	1	5	2	8	5	8	2	5	Imp
53	1	2	1	1	1	1	9	9	9	9	5	1	1	5	2	7	6	8	3	4	Imp

54	1	2	1	1	1	1	9	9	9	9	5	1	1	5	2	6	7	8	4	3	Imp
55	1	2	1	1	1	1	9	9	9	9	5	1	1	5	2	5	8	8	5	2	Imp
56	1	2	1	1	1	1	9	9	9	9	5	1	1	5	2	4	9	8	6	1	Imp
57	1	2	1	1	1	1	9	9	9	9	5	1	1	5	1	9	5	9	1	5	Imp
58	1	2	1	1	1	1	9	9	9	9	5	1	1	5	1	8	6	9	2	4	Imp
59	1	2	1	1	1	1	9	9	9	9	5	1	1	5	1	7	7	9	3	3	Imp
60	1	2	1	1	1	1	9	9	9	9	5	1	1	5	1	6	8	9	4	2	Imp
61	1	2	1	1	1	1	9	9	9	9	5	1	1	5	1	5	9	9	5	1	Imp
62	1	2	1	1	1	1	9	9	9	8	6	1	2	4	9	5	1	1	5	9	Imp
63	1	2	1	1	1	1	9	9	9	8	6	1	2	4	9	4	2	1	6	8	Imp
64	1	2	1	1	1	1	9	9	9	8	6	1	2	4	9	3	3	1	7	7	Imp
65	1	2	1	1	1	1	9	9	9	8	6	1	2	4	9	2	4	1	8	6	Imp
66	1	2	1	1	1	1	9	9	9	8	6	1	2	4	9	1	5	1	9	5	Imp
67	1	2	1	1	1	1	9	9	9	8	6	1	2	4	8	6	1	2	4	9	Imp
68	1	2	1	1	1	1	9	9	9	8	6	1	2	4	8	5	2	2	5	8	Imp
69	1	2	1	1	1	1	9	9	9	8	6	1	2	4	8	4	3	2	6	7	Imp
70	1	2	1	1	1	1	9	9	9	8	6	1	2	4	8	3	4	2	7	6	Imp
71	1	2	1	1	1	1	9	9	9	8	6	1	2	4	8	2	5	2	8	5	Imp
72	1	2	1	1	1	1	9	9	9	8	6	1	2	4	8	1	6	2	9	4	Imp
73	1	2	1	1	1	1	9	9	9	8	6	1	2	4	7	7	1	3	3	9	Imp
74	1	2	1	1	1	1	9	9	9	8	6	1	2	4	7	6	2	3	4	8	Imp
75	1	2	1	1	1	1	9	9	9	8	6	1	2	4	7	5	3	3	5	7	Imp
76	1	2	1	1	1	1	9	9	9	8	6	1	2	4	7	4	4	3	6	6	Imp
77	1	2	1	1	1	1	9	9	9	8	6	1	2	4	7	3	5	3	7	5	Imp
78	1	2	1	1	1	1	9	9	9	8	6	1	2	4	7	2	6	3	8	4	Imp
79	1	2	1	1	1	1	9	9	9	8	6	1	2	4	7	1	7	3	9	3	Imp
80	1	2	1	1	1	1	9	9	9	8	6	1	2	4	6	8	1	4	2	9	Imp

81	1	2	1	1	1	1	9	9	9	8	6	1	2	4	6	7	2	4	3	8	Imp
82	1	2	1	1	1	1	9	9	9	8	6	1	2	4	6	6	3	4	4	7	Imp
83	1	2	1	1	1	1	9	9	9	8	6	1	2	4	6	5	4	4	5	6	Imp
84	1	2	1	1	1	1	9	9	9	8	6	1	2	4	6	4	5	4	6	5	Imp
85	1	2	1	1	1	1	9	9	9	8	6	1	2	4	6	3	6	4	7	4	Imp
86	1	2	1	1	1	1	9	9	9	8	6	1	2	4	6	2	7	4	8	3	Imp
87	1	2	1	1	1	1	9	9	9	8	6	1	2	4	6	1	8	4	9	2	Imp
88	1	2	1	1	1	1	9	9	9	8	6	1	2	4	5	9	1	5	1	9	Imp
89	1	2	1	1	1	1	9	9	9	8	6	1	2	4	5	8	2	5	2	8	Imp
90	1	2	1	1	1	1	9	9	9	8	6	1	2	4	5	7	3	5	3	7	Imp
91	1	2	1	1	1	1	9	9	9	8	6	1	2	4	5	6	4	5	4	6	Imp
92	1	2	1	1	1	1	9	9	9	8	6	1	2	4	5	5	5	5	5	5	Imp
93	1	2	1	1	1	1	9	9	9	8	6	1	2	4	5	4	6	5	6	4	Imp
94	1	2	1	1	1	1	9	9	9	8	6	1	2	4	5	3	7	5	7	3	Imp
95	1	2	1	1	1	1	9	9	9	8	6	1	2	4	5	2	8	5	8	2	Imp
96	1	2	1	1	1	1	9	9	9	8	6	1	2	4	5	1	9	5	9	1	Imp
97	1	2	1	1	1	1	9	9	9	8	6	1	2	4	4	9	2	6	1	8	Imp
98	1	2	1	1	1	1	9	9	9	8	6	1	2	4	4	8	3	6	2	7	Imp
99	1	2	1	1	1	1	9	9	9	8	6	1	2	4	4	7	4	6	3	6	Imperfect
100	1	2	1	1	1	1	9	9	9	8	6	1	2	4	4	6	5	6	4	5	Imperfect

PERFECT FRAME TWO (2)
AFAMATRIX MAGIC SQUARE GRID TABLE
(Showing the first 100 out of the 1,584 Possibilities)

Perfect F2 S/N	F2 S/N	X- Component Identifier				X- Component				Plus-Component												CLASS
		X- Comp S/N	GROUP	SELECTION	POSITION	TL	TR	BR	BL	VERTICAL						HORIZONTAL						
										TOP-SIDE			DOWN-SIDE			LEFT-SIDE			RIGHT-SIDE			
						1	1'	2	2'	A	B	C	A'	B'	C'	a	b	c	a'	b'	c'	

1	381	1	2	1	1	1	1	9	9	8	8	7	2	2	3	7	4	4	3	6	6	P
2	387	1	2	1	1	1	1	9	9	8	8	7	2	2	3	6	6	3	4	4	7	P
3	390	1	2	1	1	1	1	9	9	8	8	7	2	2	3	6	3	6	4	7	4	P
4	404	1	2	1	1	1	1	9	9	8	8	7	2	2	3	4	7	4	6	3	6	P
5	407	1	2	1	1	1	1	9	9	8	8	7	2	2	3	4	4	7	6	6	3	P
6	413	1	2	1	1	1	1	9	9	8	8	7	2	2	3	3	6	6	7	4	4	P
7	442	1	2	1	1	1	1	9	9	8	7	8	2	3	2	7	4	4	3	6	6	P
8	448	1	2	1	1	1	1	9	9	8	7	8	2	3	2	6	6	3	4	4	7	P
9	451	1	2	1	1	1	1	9	9	8	7	8	2	3	2	6	3	6	4	7	4	P
10	465	1	2	1	1	1	1	9	9	8	7	8	2	3	2	4	7	4	6	3	6	P
11	468	1	2	1	1	1	1	9	9	8	7	8	2	3	2	4	4	7	6	6	3	P
12	474	1	2	1	1	1	1	9	9	8	7	8	2	3	2	3	6	6	7	4	4	P
13	625	1	2	1	1	1	1	9	9	7	8	8	3	2	2	7	4	4	3	6	6	P
14	631	1	2	1	1	1	1	9	9	7	8	8	3	2	2	6	6	3	4	4	7	P
15	634	1	2	1	1	1	1	9	9	7	8	8	3	2	2	6	3	6	4	7	4	P
16	648	1	2	1	1	1	1	9	9	7	8	8	3	2	2	4	7	4	6	3	6	P
17	651	1	2	1	1	1	1	9	9	7	8	8	3	2	2	4	4	7	6	6	3	P
18	657	1	2	1	1	1	1	9	9	7	8	8	3	2	2	3	6	6	7	4	4	P
19	2242	3	3	3	1	1	3	9	7	9	8	4	1	2	6	8	6	3	2	4	7	P
20	2245	3	3	3	1	1	3	9	7	9	8	4	1	2	6	8	3	6	2	7	4	P
21	2258	3	3	3	1	1	3	9	7	9	8	4	1	2	6	6	8	3	4	2	7	P
22	2263	3	3	3	1	1	3	9	7	9	8	4	1	2	6	6	3	8	4	7	2	P
23	2279	3	3	3	1	1	3	9	7	9	8	4	1	2	6	3	8	6	7	2	4	P
24	2281	3	3	3	1	1	3	9	7	9	8	4	1	2	6	3	6	8	7	4	2	P
25	2355	3	3	3	1	1	3	9	7	9	6	6	1	4	4	8	7	2	2	3	8	P
26	2360	3	3	3	1	1	3	9	7	9	6	6	1	4	4	8	2	7	2	8	3	P
27	2363	3	3	3	1	1	3	9	7	9	6	6	1	4	4	7	8	2	3	2	8	P
28	2369	3	3	3	1	1	3	9	7	9	6	6	1	4	4	7	2	8	3	8	2	P

29	2398	3	3	3	1	**1**	**3**	**9**	**7**	**9**	**6**	**6**	**1**	**4**	**4**	**2**	**8**	**7**	**8**	**2**	**3**	P
30	2399	3	3	3	1	**1**	**3**	**9**	**7**	**9**	**6**	**6**	**1**	**4**	**4**	**2**	**7**	**8**	**8**	**3**	**2**	P
31	2470	3	3	3	1	**1**	**3**	**9**	**7**	**9**	**4**	**8**	**1**	**6**	**2**	**8**	**6**	**3**	**2**	**4**	**7**	P
32	2473	3	3	3	1	**1**	**3**	**9**	**7**	**9**	**4**	**8**	**1**	**6**	**2**	**8**	**3**	**6**	**2**	**7**	**4**	P
33	2486	3	3	3	1	**1**	**3**	**9**	**7**	**9**	**4**	**8**	**1**	**6**	**2**	**6**	**8**	**3**	**4**	**2**	**7**	P
34	2491	3	3	3	1	**1**	**3**	**9**	**7**	**9**	**4**	**8**	**1**	**6**	**2**	**6**	**3**	**8**	**4**	**7**	**2**	P
35	2507	3	3	3	1	**1**	**3**	**9**	**7**	**9**	**4**	**8**	**1**	**6**	**2**	**3**	**8**	**6**	**7**	**2**	**4**	P
36	2509	3	3	3	1	**1**	**3**	**9**	**7**	**9**	**4**	**8**	**1**	**6**	**2**	**3**	**6**	**8**	**7**	**4**	**2**	P
37	2584	3	3	3	1	**1**	**3**	**9**	**7**	**8**	**9**	**4**	**2**	**1**	**6**	**8**	**6**	**3**	**2**	**4**	**7**	P
38	2587	3	3	3	1	**1**	**3**	**9**	**7**	**8**	**9**	**4**	**2**	**1**	**6**	**8**	**3**	**6**	**2**	**7**	**4**	P
39	2600	3	3	3	1	**1**	**3**	**9**	**7**	**8**	**9**	**4**	**2**	**1**	**6**	**6**	**8**	**3**	**4**	**2**	**7**	P
40	2605	3	3	3	1	**1**	**3**	**9**	**7**	**8**	**9**	**4**	**2**	**1**	**6**	**6**	**3**	**8**	**4**	**7**	**2**	P
41	2621	3	3	3	1	**1**	**3**	**9**	**7**	**8**	**9**	**4**	**2**	**1**	**6**	**3**	**8**	**6**	**7**	**2**	**4**	P
42	2623	3	3	3	1	**1**	**3**	**9**	**7**	**8**	**9**	**4**	**2**	**1**	**6**	**3**	**6**	**8**	**7**	**4**	**2**	P
43	2690	3	3	3	1	**1**	**3**	**9**	**7**	**8**	**7**	**6**	**2**	**3**	**4**	**9**	**6**	**2**	**1**	**4**	**8**	P
44	2694	3	3	3	1	**1**	**3**	**9**	**7**	**8**	**7**	**6**	**2**	**3**	**4**	**9**	**2**	**6**	**1**	**8**	**4**	P
45	2713	3	3	3	1	**1**	**3**	**9**	**7**	**8**	**7**	**6**	**2**	**3**	**4**	**6**	**9**	**2**	**4**	**1**	**8**	P
46	2720	3	3	3	1	**1**	**3**	**9**	**7**	**8**	**7**	**6**	**2**	**3**	**4**	**6**	**2**	**9**	**4**	**8**	**1**	P
47	2739	3	3	3	1	**1**	**3**	**9**	**7**	**8**	**7**	**6**	**2**	**3**	**4**	**2**	**9**	**6**	**8**	**1**	**4**	P
48	2742	3	3	3	1	**1**	**3**	**9**	**7**	**8**	**7**	**6**	**2**	**3**	**4**	**2**	**6**	**9**	**8**	**4**	**1**	P
49	2747	3	3	3	1	**1**	**3**	**9**	**7**	**8**	**6**	**7**	**2**	**4**	**3**	**9**	**6**	**2**	**1**	**4**	**8**	P
50	2751	3	3	3	1	**1**	**3**	**9**	**7**	**8**	**6**	**7**	**2**	**4**	**3**	**9**	**2**	**6**	**1**	**8**	**4**	P
51	2770	3	3	3	1	**1**	**3**	**9**	**7**	**8**	**6**	**7**	**2**	**4**	**3**	**6**	**9**	**2**	**4**	**1**	**8**	P
52	2777	3	3	3	1	**1**	**3**	**9**	**7**	**8**	**6**	**7**	**2**	**4**	**3**	**6**	**2**	**9**	**4**	**8**	**1**	P
53	2796	3	3	3	1	**1**	**3**	**9**	**7**	**8**	**6**	**7**	**2**	**4**	**3**	**2**	**9**	**6**	**8**	**1**	**4**	P
54	2799	3	3	3	1	**1**	**3**	**9**	**7**	**8**	**6**	**7**	**2**	**4**	**3**	**2**	**6**	**9**	**8**	**4**	**1**	P
55	2869	3	3	3	1	**1**	**3**	**9**	**7**	**8**	**4**	**9**	**2**	**6**	**1**	**8**	**6**	**3**	**2**	**4**	**7**	P
56	2872	3	3	3	1	**1**	**3**	**9**	**7**	**8**	**4**	**9**	**2**	**6**	**1**	**8**	**3**	**6**	**2**	**7**	**4**	P

57	2885	3	3	3	1	1	3	9	7	8	4	9	2	6	1	6	8	3	4	2	7	P
58	2890	3	3	3	1	1	3	9	7	8	4	9	2	6	1	6	3	8	4	7	2	P
59	2906	3	3	3	1	1	3	9	7	8	4	9	2	6	1	3	8	6	7	2	4	P
60	2908	3	3	3	1	1	3	9	7	8	4	9	2	6	1	3	6	8	7	4	2	P
61	2975	3	3	3	1	1	3	9	7	7	8	6	3	2	4	9	6	2	1	4	8	P
62	2979	3	3	3	1	1	3	9	7	7	8	6	3	2	4	9	2	6	1	8	4	P
63	2998	3	3	3	1	1	3	9	7	7	8	6	3	2	4	6	9	2	4	1	8	P
64	3005	3	3	3	1	1	3	9	7	7	8	6	3	2	4	6	2	9	4	8	1	P
65	3024	3	3	3	1	1	3	9	7	7	8	6	3	2	4	2	9	6	8	1	4	P
66	3027	3	3	3	1	1	3	9	7	7	8	6	3	2	4	2	6	9	8	4	1	P
67	3089	3	3	3	1	1	3	9	7	7	6	8	3	4	2	9	6	2	1	4	8	P
68	3093	3	3	3	1	1	3	9	7	7	6	8	3	4	2	9	2	6	1	8	4	P
69	3112	3	3	3	1	1	3	9	7	7	6	8	3	4	2	6	9	2	4	1	8	P
70	3119	3	3	3	1	1	3	9	7	7	6	8	3	4	2	6	2	9	4	8	1	P
71	3138	3	3	3	1	1	3	9	7	7	6	8	3	4	2	2	9	6	8	1	4	P
72	3141	3	3	3	1	1	3	9	7	7	6	8	3	4	2	2	6	9	8	4	1	P
73	3210	3	3	3	1	1	3	9	7	6	9	6	4	1	4	8	7	2	2	3	8	P
74	3215	3	3	3	1	1	3	9	7	6	9	6	4	1	4	8	2	7	2	8	3	P
75	3218	3	3	3	1	1	3	9	7	6	9	6	4	1	4	7	8	2	3	2	8	P
76	3224	3	3	3	1	1	3	9	7	6	9	6	4	1	4	7	2	8	3	8	2	P
77	3253	3	3	3	1	1	3	9	7	6	9	6	4	1	4	2	8	7	8	2	3	P
78	3254	3	3	3	1	1	3	9	7	6	9	6	4	1	4	2	7	8	8	3	2	P
79	3260	3	3	3	1	1	3	9	7	6	8	7	4	2	3	9	6	2	1	4	8	P
80	3264	3	3	3	1	1	3	9	7	6	8	7	4	2	3	9	2	6	1	8	4	P
81	3283	3	3	3	1	1	3	9	7	6	8	7	4	2	3	6	9	2	4	1	8	P
82	3290	3	3	3	1	1	3	9	7	6	8	7	4	2	3	6	2	9	4	8	1	P
83	3309	3	3	3	1	1	3	9	7	6	8	7	4	2	3	2	9	6	8	1	4	P
84	3312	3	3	3	1	1	3	9	7	6	8	7	4	2	3	2	6	9	8	4	1	P

85	3317	3	3	3	1	**1**	**3**	**9**	**7**	**6**	**7**	**8**	**4**	**3**	**2**	**9**	**6**	**2**	**1**	**4**	**8**	P
86	3321	3	3	3	1	**1**	**3**	**9**	**7**	**6**	**7**	**8**	**4**	**3**	**2**	**9**	**2**	**6**	**1**	**8**	**4**	P
87	3340	3	3	3	1	**1**	**3**	**9**	**7**	**6**	**7**	**8**	**4**	**3**	**2**	**6**	**9**	**2**	**4**	**1**	**8**	P
88	3347	3	3	3	1	**1**	**3**	**9**	**7**	**6**	**7**	**8**	**4**	**3**	**2**	**6**	**2**	**9**	**4**	**8**	**1**	P
89	3366	3	3	3	1	**1**	**3**	**9**	**7**	**6**	**7**	**8**	**4**	**3**	**2**	**2**	**9**	**6**	**8**	**1**	**4**	P
90	3369	3	3	3	1	**1**	**3**	**9**	**7**	**6**	**7**	**8**	**4**	**3**	**2**	**2**	**6**	**9**	**8**	**4**	**1**	P
91	3381	3	3	3	1	**1**	**3**	**9**	**7**	**6**	**6**	**9**	**4**	**4**	**1**	**8**	**7**	**2**	**2**	**3**	**8**	P
92	3386	3	3	3	1	**1**	**3**	**9**	**7**	**6**	**6**	**9**	**4**	**4**	**1**	**8**	**2**	**7**	**2**	**8**	**3**	P
93	3389	3	3	3	1	**1**	**3**	**9**	**7**	**6**	**6**	**9**	**4**	**4**	**1**	**7**	**8**	**2**	**3**	**2**	**8**	P
94	3395	3	3	3	1	**1**	**3**	**9**	**7**	**6**	**6**	**9**	**4**	**4**	**1**	**7**	**2**	**8**	**3**	**8**	**2**	P
95	3424	3	3	3	1	**1**	**3**	**9**	**7**	**6**	**6**	**9**	**4**	**4**	**1**	**2**	**8**	**7**	**8**	**2**	**3**	P
96	3425	3	3	3	1	**1**	**3**	**9**	**7**	**6**	**6**	**9**	**4**	**4**	**1**	**2**	**7**	**8**	**8**	**3**	**2**	P
97	3610	3	3	3	1	**1**	**3**	**9**	**7**	**4**	**9**	**8**	**6**	**1**	**2**	**8**	**6**	**3**	**2**	**4**	**7**	P
98	3613	3	3	3	1	**1**	**3**	**9**	**7**	**4**	**9**	**8**	**6**	**1**	**2**	**8**	**3**	**6**	**2**	**7**	**4**	P
99	3626	3	3	3	1	**1**	**3**	**9**	**7**	**4**	**9**	**8**	**6**	**1**	**2**	**6**	**8**	**3**	**4**	**2**	**7**	P
100	3631	3	3	3	1	**1**	**3**	**9**	**7**	**4**	**9**	**8**	**6**	**1**	**2**	**6**	**3**	**8**	**4**	**7**	**2**	P

5) Magic Square Grid Variations Table

For us to find all the possible variations of a Magic Square Grid, a **Magic Square Grid Variations Table** needs to be constructed. The table is made up of a **Cross-Join** of all the Frame Computation Tables of the Magic Square Grid.

Therefore, an Odd Ordered Magic Square Grid Variations Table is made up of **Square 0** Cross-Join **Frame 1 Table** Cross Join **Frame 2 Table** Cross-Join **Frame 3 Table** E.T.C

Mathematically, **Total Possible Odd Order Magic Square Grid Variations** =Number of Center Square Variations * Number of Frame 1 Variations * Number of Frame 2 Variations * Number of Frame 3 Variations E.T.C.
Likewise, an Even Ordered Magic Square Grid Variations Table is made up of **Square 0.5**Cross-Join **Frame 1.5Table** Cross Join **Frame 2.5Table** Cross-Join **Frame 3.5 Table** E.T.C

Mathematically, **Total Possible Even Order Magic Square Grid Variations** =Number of Center Square Variations*Number of Frame 1 Variations*Number of Frame 2 Variations*Number of Frame 3 Variations E.T.C.

LEVEL 1 AFAMATRIX MAGIC SQUARE GRID TABLE (Showing the Perfect and Imperfect Possibilities)

Level 1 S/N	X- Component Identifier				Square '0'	X- Component				Plus-Component				CLASS
										Vertical		Horizontal		
	X-Comp S/N	Group	Selection	Position		T L	T R	B R	B L	T. S.	D. S.	L. S.	R. S.	
						1	2	1'	2'	A	A'	A	a'	
1	5	2	5	1	5	1	5	9	5	9	1	9	1	ImP
2	13	3	8	1	5	2	4	8	6	9	1	7	3	P
3	14	2	9	1	5	2	5	8	5	8	2	8	2	ImP
4	15	3	8	5	5	2	6	8	4	7	3	9	1	P
5	21	2	10	1	5	3	3	7	7	9	1	5	5	ImP
6	22	3	11	1	5	3	4	7	6	8	2	6	4	ImP

7	23	2	12	1	5	3	5	7	5	7	3	7	3	ImP
8	24	3	11	5	5	3	6	7	4	6	4	8	2	ImP
9	25	2	10	4	5	3	7	7	3	5	5	9	1	ImP
10	29	3	8	8	5	4	2	6	8	9	1	3	7	P
11	30	3	11	8	5	4	3	6	7	8	2	4	6	ImP
12	31	2	13	1	5	4	4	6	6	7	3	5	5	ImP
13	32	2	14	1	5	4	5	6	5	6	4	6	4	ImP
14	33	2	13	4	5	4	6	6	4	5	5	7	3	ImP
15	34	3	11	4	5	4	7	6	3	4	6	8	2	ImP
16	35	3	8	4	5	4	8	6	2	3	7	9	1	P
17	37	2	5	2	5	5	1	5	9	9	1	1	9	ImP
18	38	2	9	2	5	5	2	5	8	8	2	2	8	ImP
19	39	2	12	2	5	5	3	5	7	7	3	3	7	ImP
20	40	2	14	2	5	5	4	5	6	6	4	4	6	ImP
21	41	1	15	1	5	5	5	5	5	5	5	5	5	ImP
22	42	2	14	4	5	5	6	5	4	4	6	6	4	ImP
23	43	2	12	4	5	5	7	5	3	3	7	7	3	ImP
24	44	2	9	4	5	5	8	5	2	2	8	8	2	ImP
25	45	2	5	4	5	5	9	5	1	1	9	9	1	ImP
26	47	3	8	2	5	6	2	4	8	7	3	1	9	P
27	48	3	11	2	5	6	3	4	7	6	4	2	8	ImP
28	49	2	13	2	5	6	4	4	6	5	5	3	7	ImP
29	50	2	14	3	5	6	5	4	5	4	6	4	6	ImP
30	51	2	13	3	5	6	6	4	4	3	7	5	5	ImP
31	52	3	11	6	5	6	7	4	3	2	8	6	4	ImP
32	53	3	8	6	5	6	8	4	2	1	9	7	3	P
33	57	2	10	2	5	7	3	3	7	5	5	1	9	ImP
34	58	3	11	7	5	7	4	3	6	4	6	2	8	ImP

35	**59**	2	12	3	**5**	**7**	**5**	**3**	**5**	**3**	**7**	**3**	**7**	ImP
36	**60**	3	11	3	**5**	**7**	**6**	**3**	**4**	**2**	**8**	**4**	**6**	ImP
37	**61**	2	10	3	**5**	**7**	**7**	**3**	**3**	**1**	**9**	**5**	**5**	ImP
38	**67**	3	8	7	**5**	**8**	**4**	**2**	**6**	**3**	**7**	**1**	**9**	P
39	**68**	2	9	3	**5**	**8**	**5**	**2**	**5**	**2**	**8**	**2**	**8**	ImP
40	**69**	3	8	3	**5**	**8**	**6**	**2**	**4**	**1**	**9**	**3**	**7**	P
41	**77**	2	5	3	**5**	**9**	**5**	**1**	**5**	**1**	**9**	**1**	**9**	ImP

LEVEL ONE (1) PERFECT AFAMATRIX MAGIC SQUARE GRID TABLE
(Showing only the Perfect Possibilities)

Perfect L1 S/N	Level 1 S/N	X- Component Identifier				Square '0'	X- Component				Plus-Component				CLASS
											VERTICAL		HORIZONTAL		
		X-Comp S/N	Group	Selection	Position		T L	T R	B R	B L	T. S.	D. S.	L. S.	R. S.	
							1	2	1'	2'	A	A'	A	a'	
1	2	**13**	3	5	1	**5**	**2**	**4**	**8**	**6**	**9**	**1**	**7**	**3**	P
2	4	**15**	3	5	5	**5**	**2**	**6**	**8**	**4**	**7**	**3**	**9**	**1**	P
3	10	**29**	3	5	8	**5**	**4**	**2**	**6**	**8**	**9**	**1**	**3**	**7**	P
4	16	**35**	3	5	4	**5**	**4**	**8**	**6**	**2**	**3**	**7**	**9**	**1**	P
5	26	**47**	3	5	2	**5**	**6**	**2**	**4**	**8**	**7**	**3**	**1**	**9**	P
6	32	**53**	3	5	6	**5**	**6**	**8**	**4**	**2**	**1**	**9**	**7**	**3**	P
7	38	**67**	3	5	7	**5**	**8**	**4**	**2**	**6**	**3**	**7**	**1**	**9**	P
8	40	**69**	3	5	3	**5**	**8**	**6**	**2**	**4**	**1**	**9**	**3**	**7**	P

PERFECT LEVEL 2 AFAMATRIX MAGIC SQUARE VARIATION TABLE
(Showing the first 100 out of the 12,678 Possibilities)

L2 Perfect Computation S/N	Frame (Square) '0'	Frame 1								Frame 2																CLASS
		X-Frame Subcomponent				Plus-Component				X-Frame Subcomponent				Plus-Component												
						VRTCL		HRZNTL						VERTICAL						HORIZONTAL						
		T L	T R	B R	B L	T. S.	D. S.	L. S.	R. S.	T L	T R	B R	B L	TOP-SIDE			DOWN-SIDE			LEFT-SIDE			RIGHT-SIDE			
		1	2	1'	2'	A	A'	a	a'	1	2	1'	2'	A	B	C	A'	B'	C'	A	b	c	a'	b'	c'	
1	5	2	4	8	6	9	1	7	3	1	7	9	3	9	7	1	1	3	9	9	9	3	1	1	7	P
2	5	2	4	8	6	9	1	7	3	1	7	9	3	9	7	1	1	3	9	9	8	4	1	2	6	P
3	5	2	4	8	6	9	1	7	3	1	7	9	3	9	7	1	1	3	9	9	7	5	1	3	5	P
4	5	2	4	8	6	9	1	7	3	1	7	9	3	9	7	1	1	3	9	9	6	6	1	4	4	P
5	5	2	4	8	6	9	1	7	3	1	7	9	3	9	7	1	1	3	9	9	5	7	1	5	3	P
6	5	2	4	8	6	9	1	7	3	1	7	9	3	9	7	1	1	3	9	9	4	8	1	6	2	P
7	5	2	4	8	6	9	1	7	3	1	7	9	3	9	7	1	1	3	9	9	3	9	1	7	1	P
8	5	2	4	8	6	9	1	7	3	1	7	9	3	9	7	1	1	3	9	8	9	4	2	1	6	P
9	5	2	4	8	6	9	1	7	3	1	7	9	3	9	7	1	1	3	9	8	8	5	2	2	5	P
10	5	2	4	8	6	9	1	7	3	1	7	9	3	9	7	1	1	3	9	8	7	6	2	3	4	P
11	5	2	4	8	6	9	1	7	3	1	7	9	3	9	7	1	1	3	9	8	6	7	2	4	3	P
12	5	2	4	8	6	9	1	7	3	1	7	9	3	9	7	1	1	3	9	8	5	8	2	5	2	P
13	5	2	4	8	6	9	1	7	3	1	7	9	3	9	7	1	1	3	9	8	4	9	2	6	1	P
14	5	2	4	8	6	9	1	7	3	1	7	9	3	9	7	1	1	3	9	7	9	5	3	1	5	P
15	5	2	4	8	6	9	1	7	3	1	7	9	3	9	7	1	1	3	9	7	8	6	3	2	4	P
16	5	2	4	8	6	9	1	7	3	1	7	9	3	9	7	1	1	3	9	7	7	7	3	3	3	P
17	5	2	4	8	6	9	1	7	3	1	7	9	3	9	7	1	1	3	9	7	6	8	3	4	2	P
18	5	2	4	8	6	9	1	7	3	1	7	9	3	9	7	1	1	3	9	7	5	9	3	5	1	P

19	5	2	4	8	6	9	1	7	3	1	7	9	3	9	7	1	1	3	9	6	9	6	4	1	4	P
20	5	2	4	8	6	9	1	7	3	1	7	9	3	9	7	1	1	3	9	6	8	7	4	2	3	P
21	5	2	4	8	6	9	1	7	3	1	7	9	3	9	7	1	1	3	9	6	7	8	4	3	2	P
22	5	2	4	8	6	9	1	7	3	1	7	9	3	9	7	1	1	3	9	6	6	9	4	4	1	P
23	5	2	4	8	6	9	1	7	3	1	7	9	3	9	7	1	1	3	9	5	9	7	5	1	3	P
24	5	2	4	8	6	9	1	7	3	1	7	9	3	9	7	1	1	3	9	5	8	8	5	2	2	P
25	5	2	4	8	6	9	1	7	3	1	7	9	3	9	7	1	1	3	9	5	7	9	5	3	1	P
26	5	2	4	8	6	9	1	7	3	1	7	9	3	9	7	1	1	3	9	4	9	8	6	1	2	P
27	5	2	4	8	6	9	1	7	3	1	7	9	3	9	7	1	1	3	9	4	8	9	6	2	1	P
28	5	2	4	8	6	9	1	7	3	1	7	9	3	9	7	1	1	3	9	3	9	9	7	1	1	P
29	5	2	4	8	6	9	1	7	3	1	7	9	3	9	6	2	1	4	8	9	9	3	1	1	7	P
30	5	2	4	8	6	9	1	7	3	1	7	9	3	9	6	2	1	4	8	9	8	4	1	2	6	P
31	5	2	4	8	6	9	1	7	3	1	7	9	3	9	6	2	1	4	8	9	7	5	1	3	5	P
32	5	2	4	8	6	9	1	7	3	1	7	9	3	9	6	2	1	4	8	9	6	6	1	4	4	P
33	5	2	4	8	6	9	1	7	3	1	7	9	3	9	6	2	1	4	8	9	5	7	1	5	3	P
34	5	2	4	8	6	9	1	7	3	1	7	9	3	9	6	2	1	4	8	9	4	8	1	6	2	P
35	5	2	4	8	6	9	1	7	3	1	7	9	3	9	6	2	1	4	8	9	3	9	1	7	1	P
36	5	2	4	8	6	9	1	7	3	1	7	9	3	9	6	2	1	4	8	8	9	4	2	1	6	P
37	5	2	4	8	6	9	1	7	3	1	7	9	3	9	6	2	1	4	8	8	8	5	2	2	5	P
38	5	2	4	8	6	9	1	7	3	1	7	9	3	9	6	2	1	4	8	8	7	6	2	3	4	P
39	5	2	4	8	6	9	1	7	3	1	7	9	3	9	6	2	1	4	8	8	6	7	2	4	3	P
40	5	2	4	8	6	9	1	7	3	1	7	9	3	9	6	2	1	4	8	8	5	8	2	5	2	P
41	5	2	4	8	6	9	1	7	3	1	7	9	3	9	6	2	1	4	8	8	4	9	2	6	1	P
42	5	2	4	8	6	9	1	7	3	1	7	9	3	9	6	2	1	4	8	7	9	5	3	1	5	P
43	5	2	4	8	6	9	1	7	3	1	7	9	3	9	6	2	1	4	8	7	8	6	3	2	4	P
44	5	2	4	8	6	9	1	7	3	1	7	9	3	9	6	2	1	4	8	7	7	7	3	3	3	P

45	5	2	4	8	6	9	1	7	3	1	7	9	3	9	6	2	1	4	8	7	6	8	3	4	2	P
46	5	2	4	8	6	9	1	7	3	1	7	9	3	9	6	2	1	4	8	7	5	9	3	5	1	P
47	5	2	4	8	6	9	1	7	3	1	7	9	3	9	6	2	1	4	8	6	9	6	4	1	4	P
48	5	2	4	8	6	9	1	7	3	1	7	9	3	9	6	2	1	4	8	6	8	7	4	2	3	P
49	5	2	4	8	6	9	1	7	3	1	7	9	3	9	6	2	1	4	8	6	7	8	4	3	2	P
50	5	2	4	8	6	9	1	7	3	1	7	9	3	9	6	2	1	4	8	6	6	9	4	4	1	P
51	5	2	4	8	6	9	1	7	3	1	7	9	3	9	6	2	1	4	8	5	9	7	5	1	3	P
52	5	2	4	8	6	9	1	7	3	1	7	9	3	9	6	2	1	4	8	5	8	8	5	2	2	P
53	5	2	4	8	6	9	1	7	3	1	7	9	3	9	6	2	1	4	8	5	7	9	5	3	1	P
54	5	2	4	8	6	9	1	7	3	1	7	9	3	9	6	2	1	4	8	4	9	8	6	1	2	P
55	5	2	4	8	6	9	1	7	3	1	7	9	3	9	6	2	1	4	8	4	8	9	6	2	1	P
56	5	2	4	8	6	9	1	7	3	1	7	9	3	9	6	2	1	4	8	3	9	9	7	1	1	P
57	5	2	4	8	6	9	1	7	3	1	7	9	3	9	5	3	1	5	7	9	9	3	1	1	7	P
58	5	2	4	8	6	9	1	7	3	1	7	9	3	9	5	3	1	5	7	9	8	4	1	2	6	P
59	5	2	4	8	6	9	1	7	3	1	7	9	3	9	5	3	1	5	7	9	7	5	1	3	5	P
60	5	2	4	8	6	9	1	7	3	1	7	9	3	9	5	3	1	5	7	9	6	6	1	4	4	P
61	5	2	4	8	6	9	1	7	3	1	7	9	3	9	5	3	1	5	7	9	5	7	1	5	3	P
62	5	2	4	8	6	9	1	7	3	1	7	9	3	9	5	3	1	5	7	9	4	8	1	6	2	P
63	5	2	4	8	6	9	1	7	3	1	7	9	3	9	5	3	1	5	7	9	3	9	1	7	1	P
64	5	2	4	8	6	9	1	7	3	1	7	9	3	9	5	3	1	5	7	8	9	4	2	1	6	P
65	5	2	4	8	6	9	1	7	3	1	7	9	3	9	5	3	1	5	7	8	8	5	2	2	5	P
66	5	2	4	8	6	9	1	7	3	1	7	9	3	9	5	3	1	5	7	8	7	6	2	3	4	P
67	5	2	4	8	6	9	1	7	3	1	7	9	3	9	5	3	1	5	7	8	6	7	2	4	3	P
68	5	2	4	8	6	9	1	7	3	1	7	9	3	9	5	3	1	5	7	8	5	8	2	5	2	P
69	5	2	4	8	6	9	1	7	3	1	7	9	3	9	5	3	1	5	7	8	4	9	2	6	1	P
70	5	2	4	8	6	9	1	7	3	1	7	9	3	9	5	3	1	5	7	7	9	5	3	1	5	P

71	5	2	4	8	6	9	1	7	3	1	7	9	3	9	5	3	1	5	7	7	8	6	3	2	4	P
72	5	2	4	8	6	9	1	7	3	1	7	9	3	9	5	3	1	5	7	7	7	7	3	3	3	P
73	5	2	4	8	6	9	1	7	3	1	7	9	3	9	5	3	1	5	7	7	6	8	3	4	2	P
74	5	2	4	8	6	9	1	7	3	1	7	9	3	9	5	3	1	5	7	7	5	9	3	5	1	P
75	5	2	4	8	6	9	1	7	3	1	7	9	3	9	5	3	1	5	7	6	9	6	4	1	4	P
76	5	2	4	8	6	9	1	7	3	1	7	9	3	9	5	3	1	5	7	6	8	7	4	2	3	P
77	5	2	4	8	6	9	1	7	3	1	7	9	3	9	5	3	1	5	7	6	7	8	4	3	2	P
78	5	2	4	8	6	9	1	7	3	1	7	9	3	9	5	3	1	5	7	6	6	9	4	4	1	P
79	5	2	4	8	6	9	1	7	3	1	7	9	3	9	5	3	1	5	7	5	9	7	5	1	3	P
80	5	2	4	8	6	9	1	7	3	1	7	9	3	9	5	3	1	5	7	5	8	8	5	2	2	P
81	5	2	4	8	6	9	1	7	3	1	7	9	3	9	5	3	1	5	7	5	7	9	5	3	1	P
82	5	2	4	8	6	9	1	7	3	1	7	9	3	9	5	3	1	5	7	4	9	8	6	1	2	P
83	5	2	4	8	6	9	1	7	3	1	7	9	3	9	5	3	1	5	7	4	8	9	6	2	1	P
84	5	2	4	8	6	9	1	7	3	1	7	9	3	9	5	3	1	5	7	3	9	9	7	1	1	P
85	5	2	4	8	6	9	1	7	3	1	7	9	3	9	4	4	1	6	6	9	9	3	1	1	7	P
86	5	2	4	8	6	9	1	7	3	1	7	9	3	9	4	4	1	6	6	9	8	4	1	2	6	P
87	5	2	4	8	6	9	1	7	3	1	7	9	3	9	4	4	1	6	6	9	7	5	1	3	5	P
88	5	2	4	8	6	9	1	7	3	1	7	9	3	9	4	4	1	6	6	9	6	6	1	4	4	P
89	5	2	4	8	6	9	1	7	3	1	7	9	3	9	4	4	1	6	6	9	5	7	1	5	3	P
90	5	2	4	8	6	9	1	7	3	1	7	9	3	9	4	4	1	6	6	9	4	8	1	6	2	P
91	5	2	4	8	6	9	1	7	3	1	7	9	3	9	4	4	1	6	6	9	3	9	1	7	1	P
92	5	2	4	8	6	9	1	7	3	1	7	9	3	9	4	4	1	6	6	8	9	4	2	1	6	P
93	5	2	4	8	6	9	1	7	3	1	7	9	3	9	4	4	1	6	6	8	8	5	2	2	5	P
94	5	2	4	8	6	9	1	7	3	1	7	9	3	9	4	4	1	6	6	8	7	6	2	3	4	P
95	5	2	4	8	6	9	1	7	3	1	7	9	3	9	4	4	1	6	6	8	6	7	2	4	3	P
96	5	2	4	8	6	9	1	7	3	1	7	9	3	9	4	4	1	6	6	8	5	8	2	5	2	P

97	5	2	4	8	6	9	1	7	3	1	7	9	3	9	4	4	1	6	6	8	4	9	2	6	1	P
98	5	2	4	8	6	9	1	7	3	1	7	9	3	9	4	4	1	6	6	7	9	5	3	1	5	P
99	5	2	4	8	6	9	1	7	3	1	7	9	3	9	4	4	1	6	6	7	8	6	3	2	4	P
100	5	2	4	8	6	9	1	7	3	1	7	9	3	9	4	4	1	6	6	7	7	7	3	3	3	P

Summary of Steps in Constructing Magic Square Grids using the Afam's Method.

STEP 1) Prepare the Comprehensive Magic Square Grids Parameters Table according to specifications.

STEP 2) For Odd Order Magic Square Grids, Chose your Single Center-Square Number.

For Even Order Magic Square Grids, Prepare your Centre-Square Table, together with the optional Identifier Table.

STEP 3) For Odd Order Magic Square Grids, Prepare your X-Frame Sub Components Table of the first Frame.

For Even Order Magic Square Grids, prepare the X-Frame Subcomponent Table of the first Frame, if different from the Centre-Square Table.

STEP 4) Prepare the Integer Compositions Selection Tables of first Frame starting from the Lower Limit Table to the upper Limit Table, as determined from the X-Frame sub Component number Range of Step 3.

Note that in the case of Frame 1, only one number is involved so there is no need generating the Integer Compositions, we only have to check and use the matching number.

STEP 5) Using the X-Frame Sub Component Table of Step 3 in relation with the Integer Compositions selection Tables of Step 4, Prepare the 1stFrame Computation Table.

STEP 6) For Odd Order Magic Square Grids, use the Center Square Number of Step 2 together with the 1st Frame Computation Table of Step 5 to Prepare the Magic Square Grid Variations Table of Order 3.

For Even Order Magic Square Grids, use the Center Square Table of Step 2 together with the 1st Frame Computation Table of Step 5 to Prepare the Magic Square Grid Variations Table of Order 4.

STEP 7) For Odd Order Magic Square Grids, Transfer the numbers from the Magic Square Grid Variations Table to the Magic Square Grid of Order 3.

For Even Order Magic Square Grids, Transfer the numbers from the Magic Square Grid Variations Table to the Magic Square Grid of Order 4.

End!

STEP 8) If Order 5 or Order 6 Magic Square Grid is specified, Repeat Step 2 to Step 6 to form the 2nd Frame after Substituting 2nd Frame for 1st Frame in the Steps, however, Prepare your X-Frame Sub Components Table of 2nd Frame if different from that of 1st Frame and the Centre-Square Table. Remember that the Integer Compositions Selection Tables will also be different at Step 4.

STEP 9) Using the Magic Square Grid Variations Table of Step 6 and 2nd Frame Variations Table derived from Step 7,prepare the Magic Square Grid Variations Table of Order 5 or Order 6.

STEP 10) Transfer the numbers from the Magic Square Grid Variations Table to the Magic Square Grid of Order 5 or Order 6.

To form Magic Square Grids of higher Orders, continue repeating Step 8 to 10 substituting the Order, the Frame

Number, and Preparing your X-Frame subcomponent Table if specified differently, and Integer Compositions selection Table according to the specified Order.

CHAPTER 9

The Mathematics of Afamatrix

Having known what Afamatrix Magic Square Grids are all about and how they are formed, I will like to present to you the mathematical relationship existing within the Afamatrix Magic Square Grids.

We will first look at the summary of the formulae representing the various relationships existing within the numbers of Afamatrix Magic Square Grids, then we will consider the details of how the formulae are deduced.

Summary of Derived Formulae

If.......

M➔ Magic Constant
n➔ Order
N➔ Number of squares in the entire grid
L➔ Level
nf➔ Total number of each integer used in a frame
S➔ Sum total of each integer used in the grid.

Then........

1) The Magic Constant (M) of Afamatrix Magic Square Grids is known to be;

$$M = 5n \qquad (1)$$

$$M = 5\sqrt{N} \qquad (2)$$

$$M = 10L + 5 \qquad (3)$$

$M = 5 \sqrt{8S + 1}$ .... (4)

2) Order (n) of Afamatrix Magic Square Grid;

$n=M/5$ .... (5)

$n = \sqrt{N}$ .... (7)

$n = 2L + 1$ .... (8)

$n = \sqrt{8S + 1}$ .... (9)

3) Total Number of Squares (*N*) in an Afamatrix Magic Square Grid can be deduced as follows;

$N = M^2/5^2$ .... (10)

$N = (M/5)^2$ .... (11)

$N = n^2$ .... (12)

$N = (L2 + 1)^2$ .... (13)

$N = 8S + 1$ .... (14)

4) The Level (L) of an Afamatrix Magic Square Grid:

$L = (n-1)/2$

$L = M-5 / 10$ .... (15)

$L = 5^{n-5} / 10 = n-1/2$ .... (16)

$= (5\sqrt{N} - 5) / 10 = (\sqrt{N} - 1) / 2$ .... (17)

$L = (2S - n_f) / nf$ .... (18)

Total number of each integer used in the grid

$$S = \frac{n_f}{2} (2 \times 1 + (n_f - 1) 1 \qquad (19)$$

$$S = \frac{n_f}{2} (2 + (n_f - 1) \qquad (20)$$

$$S = \frac{n_f}{2} (1 + L) \qquad (21)$$

$$S = \frac{N - 1}{8} \qquad (22)$$

Derivation of Afamatrix Magic Square Grid Formulae.

1) The Magic Constant (M) of Afamatrix Magic Square Grids is known to be;

$$\mathbf{M=5n} \quad \ldots\ldots\ldots\ldots\ldots\ldots\ldots\ldots\ldots \quad (1)$$

Where ***'n'*** is the order or number of squares in a row, column or diagonal.

2) Deduction of the Order (n) of Afamatrix Magic Square Grid;

By *change of subject of formula from the above expression,* we will get:

$$\mathbf{n = \frac{M}{5}} \quad \ldots\ldots\ldots\ldots\ldots\ldots\ldots\ldots \quad (2)$$

3) Total Number of Squares (*N*) in an Afamatrix Magic Square Grid can be deduced as follows;
If '*N*' should represent the total number of squares in the entire grid, and the number of rows and columns being equal, then:

$$\mathbf{N = n^2} \quad \ldots\ldots\ldots\ldots\ldots\ldots\ldots\ldots \quad (3)$$

Therefore; $\mathbf{n = \sqrt{N}}$ .... (4)

From formula (1) above, therefore:

$$\mathbf{M = 5\sqrt{N}} \ldots\ldots\ldots\ldots\ldots\ldots\ldots\ldots \quad (5)$$

By *change of subject of formula* we will get:

$$\mathbf{5\sqrt{N} = M}$$

$$N = \frac{M}{5}$$

$$N = \frac{M^2}{5^2} \qquad \text{.......................} (6)$$

4) The Level (L) of an Afamatrix Magic Square Grid:

Furthermore, If L = Level, representing the size of grid, where the smallest non-trivial case in Afamatrix Magic square Grid = Order 3 = Level one (L1) = 3 x 3 grid
i.e:

Level One (L0) = Order (n) 1= 1 x 1 Square (Trivial)
Level One and Half (L1.5) = Order (n) 2= 2 x 2 grid (Trivial)

Level One (L1) = Order (n) 3 = 3 x 3 grid
Level Two (L2) = Order (n) 5 = 5 x 5 grid
Level Three (L3) = Order (n) 7 = 7 x 7 grid
Level Four (L4) = Order (n) 9 = 9 x 9 grid
Level Five (L5) = Order (n) 11 = 11 x 11 grid

And so on!

The relationship between the Level (L) and the Order (n) of Afamatrix Magic Square Grid has been found to be:

L = n-1 /2

In any selected frame of an Afamatrix grid, the numbers in the opposite squares are complimentary i.e. they add up to 10.

Therefore; if L = level

Then; the Magic constant is the product of 10 and the Level plus 5.

$$\mathbf{M = 10L + 5} \text{ .... (M3)}$$

By *change of subject of formula,* we will get:

$$\mathbf{L = \frac{M-5}{10}} \text{ ....(L1)}$$

(Please note the difference between the formula No 'L1' and the Afamatrix grid size Level 'L1')

Having seen that: M = 5n as shown in formula (M1) above, We will substitute M for 5n. Therefore:

$$\mathbf{L = \frac{5n-5}{10} = \frac{n-1}{2}} \text{ ....(L2)}$$

By *Change of subject of formula* in formula L2 above, we will get:

$$\mathbf{L \times 2 = n-1}$$

Therefore: $\mathbf{n = 2L + 1}$ .... (n3)

Also, since $M = 5\sqrt{N}$ as shown in formula (M2) above, We substitute M for $5\sqrt{N}$, Therefore:

$$\mathbf{L = \frac{5\sqrt{N} - 5}{10} = \frac{\sqrt{N} - 1}{2}} \text{ .... (L3)}$$

By *Change of subject of formula,* we will get:

$$L \times 2 = \sqrt{N} - 1$$

$$\sqrt{N} = L2 + 1$$

Therefore the relationship between the total number of squares inside the grid (N) and the Level (L) and can be deduced as:

$$\mathbf{N = (L2 + 1)^2} \text{........................ (N3)}$$

Statistics

How many digits does it take to fill an Afamatrix Magic Square Grid?

The number of each digit used in each odd numbered level of the first 21Afamatrixgrid is illustrated in the table below, With the exception of digit 5 which is always used once in each level.

LEVEL (L)	**TYPE OF GRID**	**NO OF EACH INTEGER USED IN THE OUTERMOST FRAME OF SQUARES.** (n_f)	**SUM TOTAL OF EACH INTEGER USED IN THE ENTIRE GRID (S)**
0	1 x 1	0	0
0.5	2 x 2		
1	3 x 3	1	1
1.5	4 x 4		
2	5 x 5	2	3
2.5	6 x 6		
3	7 x 7	3	6
3.5	8 x 8		
4	9 x 9	4	10
4.5	10 x 10		
5	11 x 11	5	15
5.5	12 x 12		

6	13 x 13	6	21
6.5	14 x 14		
7	15 x 15	7	28
7.5	16 x 16		
8	17 X 17	8	36
8.5	18 x 18		
9	19 X 19	9	45
9.5	20 x 20		
10	21 X 21	10	55
10.5	22 x 22		

The figures continue uniformly till any level of a perfect Afamatrix Magic Square Grid.

We can see from the table that the number of each integer used in the outermost frame of squares per level increases gradually as the level increases with a difference of +1, and with each level having a total sum that is uniform as well. Therefore we can easily conclude that the relationship between the Level (L), The number of each integer used in the outermost frame of squares (n_f) and the sum total of each integer used in the entire grid (S), forms an arithmetic progression with '1' as the first number (a), and '1' as the common difference (d),

The Level (L) is equivalent to the Last term (L) of the arithmetic progression.

Also note that the sum (S) of the arithmetic progression is also equal the total number of each integer used in the entire grid.

The sum of an arithmetic progression is given by,

$$S = n/2\ (2a + (n - 1)\ d$$

OR

$$S = n/2\,(a + L)$$

For the sake of Afamatrix, I represented n with n_f, in order to avoid confusing it with the n that denotes the number of squares per row column or diagonal.
For every increase in the level of a perfect Afamatrix grid;

a= d = 1

Therefore;

The Arithmetic progression sum of an Afamatrix grid which represents the sum total of each integer used in the entire grid; is given by:

I.e. $$S = \frac{n_f}{2}\,(2 + (n_f - 1)) \qquad \ldots\ldots\ldots\ldots\ldots\ (S1)$$

$$S = \frac{n_f}{2}\,(2 \times 1 + (n_f - 1)\,1$$

OR

Applying the 2nd formula of the sum of an arithmetic progression:

$$S = \frac{n_f}{2}(1 + L) \qquad \ldots\ldots\ldots\ldots\ldots\ldots\ldots\ (S2)$$

Likewise, the relationship between the Sum total number of each integer used in the entire grid (S) and the Total number of squares (N) is given by:

$$\mathbf{S = \frac{N - 1}{8}} \quad \ldots\ldots\ldots\ldots\ldots\ldots\ldots\ldots (S3)$$

By change of Subject of formula, we get:

$N - 1 = 8S$

Therefore: $\mathbf{N = 8S + 1}$ ....(N4)

From equation (N1) : $\mathbf{N = n^2}$

Therefore: $\mathbf{n^2 = 8S + 1}$

$\mathbf{n = \sqrt{8S + 1}}$**..........................** (n4)

Likewise: from equation (N2) above: $N = (M/5)^2$

Therefore: $(M/5)^2 = 8S + 1$

$M/5 = \sqrt{8S + 1}$

Therefore: $\mathbf{M = 5\sqrt{8S + 1}}$ ....(M4)

<u>**TOTAL POSSIBLE VARIATIONS OF AN AFAMATRIX MAGIC SQUARE GRID.**</u>

The variations in Afamatrix represent a combination of the total possible component manipulations, Classes, and formations of a particular level or Order of an Afamatrix Magic Square Grid.

Table of Total Possible Perfect Variations of Afamatrix Magic Square Grid from Level 1 to Level 10 (Order 3 to Order 21)

Level L	Order n	Grid Size n^2	Total Possible Variations TPV
1	3	3 X 3	8
2	5	5 X 5	4608
3	7	7 X 7	398,13 1,200
4	9	9 X 9	1.6855 3E+15
5	11	11X11	3.0826 9E+22
6	13	13 X 13	3.8673 2E+30
7	15	15 X 15	6.2633 6E+39
8	17	17 X 17	3.6970 7E+49
9	19	19 X 19	3.9202 8E+60
10	21	21 X 21	1.1025 3E+72

NOTE: Perfect Afamatrix Magic Square Grids are not possible in grids of even numbered Order.

Table of details of the Total Possible Perfect Variations of Afamatrix Magic Square Grid from Level 1 to Level 10 (Order 3 to Order 21)

Level L	Order n	Grid Size n^2	Number Of Perfect Forms in Outermost Frame PF	Basic Grid Positions Of Outermost Frame BGP	Number of Permutable Squares in the Vertical or Horizontal Squares of the Outermost Frame n-2	Total Number of Vertical Permutations in Outermost Frame VP	Total Number of Horizontal Permutations In Outermost Frame HP	Total Possible Variations in Outermost Frame TPVOF	Total Possible Variations In the entire Level TPV
						= P(Vn-2), K_{Vn-2} = (Vn-2)!	= P(Hn-2), K_{Hn-2} = (Hn-2)!	= PF * BGP * VP X HP	= TPVOF$_{L1}$ * TPVOF$_{L2}$ * TPVOF$_{L...?..}$
1	3	3 X 3	1	8	1	1	1	8	8
2	5	5 X 5	2	8	3	6	6	576	4608
3	7	7 X 7	6	8	5	120	120	691200	398,131,200
4	9	9 X 9	12	8	7	5040	5040	2438553600	1.68553E+15
5	11	11X 11	12	8	9	362880	362880	1.26415E+13	3.08269E+22
6	13	13 X 13	24	8	11	39916800	39916800	3.05923E+17	3.86732E+30
7	15	15 X 15	66	8	13	6227020800	6227020800	2.04736E+22	6.26336E+39
8	17	17 X 17	132	8	15	1.30767E+12	1.30767E+12	1.80577E+27	3.69707E+49
9	19	19 X 19	2145	8	17	3.55687E+14	3.55687E+14	2.17097E+33	3.92028E+60
10	21	21 X 21	4290	8	19	1.21645E+17	1.21645E+17	5.07851E+38	1.10253E+72

CHAPTER 10

Afamatirix Magic Square Grid Puzzles:

One of the Practical Applications of Afamatrix Magic Square Grids.

One of the most important practical applications of Afamatrix Magic Square Grids could be its easy adaptation to puzzles of endless Combinations.
When applied as a puzzle by omitting some numbers in some boxes of the grid, the number arrangements of a particular unique Afamatrix Magic Square Grid can be combined to form a different puzzle.

The Difficulty Sub-Level Combinations of Afamatrix Puzzles.

The Table below shows the Difficulty Sub-Level Combinations of Frame 1 to frame 4 of Afamatrix Magic Square Grid Puzzles, and the number of non-blank cells that are available on each frame.

The Combination notation represents the different possible sub-level number arrangements in a given frame of an Afamatrix puzzle. Only the combination involving non-complimentary Numbers are accepted, others are considered as exceptions. The colors represent the sub-levels that will be used in the inner frames of a given sub-level in each Level of Afamatrix Magic Square Grid.
The Number of Non-Blank (NB) Squares that forms the hardest Sub-Level of a given Frame is determined by the ability to get only one unique Afamatrix Magic Square Grid if the Blank Squares (BS) are to be filled.

				Level			
				← L4 →			
				← L3 →			
				←L2→			
				L1			
				F1	F2	F3	F4
Sub–Level	SL1	VE	E	4	8	12	16
	SL2	E			7	11	15
	SL3	I	I	3	6	10	14
	SL4	H	H	2	5	9	13
	SL5	VH			4	8	12

SL1	>	VE	>	VERY EASY
SL2	>	E	>	EASY
SL3	>	I	>	INTERMEDIATE
SL4	>	H	>	HARD
SL5	>	VH	>	VERY HARD

Total Squares (TS) =Non-Blank Squares (NBS) + Blank Squares (BS)

Total Squares (TS) are the total number of Squares in a Frame.

Non-Blank Squares or Chosen Squares(NBS) are the pre-filled Squares in a Frame, used in determining the difficulty sublevel of the Afamatrix puzzle
Blank Squares (BS)are the empty Squares in a frame, used in determining the difficulty Sub-Level of the Afamatrix puzzle.

Total Combinations (TC)are the total possible combinations that can be formed by the Mathematical Combination of Total Squares (TS) and the Chosen Squares (NBS)

Valid Combinations(VC)are the Combinations that that can be used in forming one unique Afamatrix Puzzle only. Such

Combinations does not allow for the formation of more than one unique Afamatrix Magic Square Grid.

Exceptions (e) are the Combinations that allows for the formation of more than one unique Afamatrix Magic Square Grid. i.e. Combinations that does not allow for the formation of a unique Afamatrix Puzzle.

Afamatrix Equilibrium Table Mechanical Puzzle

An Afamatrix Equilibrium Table Mechanical Puzzle can be constructed by cutting out solid shapes of different sizes analogues to the numbers in an Afamatrix Magic Square Grid. The shapes are arranged on the flat surface of a square shaped board which is rested on a pointed fulcrum.

The idea is to arrange the shapes on the board such that when it is placed on the fulcrum, the board will maintain balance, therefore not tipping on one side.

Afamatrix Code

Afamatrix is made of Magic Square Grids of endless variation which can be turned into Magic Square Grid Puzzles of endless Combinations. Therefore, There is a need to create standardized simple way of representing the endless variations and combinations. Hence, The Afamatrix Code.

Afamatrix Code is made of two parts namely the Variation Code and the Combination Code.

For instance: Afamatrix ST.L1:1-SL2:89

Afamatrix Magic Square Grid Variation Code

The Afamatrix Variation codes are the set of alphanumeric notation representing the various Classes, Levels and Serial Number of Selected Afamatrix Magic Square Grids from the Authorized Magic Square Variation Tables. Thus, every unique Afamatrix Magic Square Gridhasan Afamatrix Variation code.

The Afamatrix Combination Codes on the other hand, are the set of alphanumeric notation representing the various Difficulty Sub-Levels and Serial Number of Selected Afamatrix Magic Square Grid Puzzles from the Authorized Puzzle Combination Tables. Thus, every unique Afamatrix Magic Square Grid Puzzle has an Afamatrix Combination Code.

The Afamatrix Variation Codes makes the construction of Afamatrix Magic Square Grids easy. This is because the codes contain information on the Class, Level and Serial Number of Selected Afamatrix Magic Square Grids. Just looking at the code, one can easily know the required Magic Square Grid, which can only be selected from the specified Table.

For Instance, the Level 1, Level 2 and Level 3 Afamatrix Magic Square Grid variations can be represented as shown below:

Afamatrix ST.L1:1-SL2:89
(Simple Table.Level1.:Serial Number 1)

AfamatrixPST.L1:1
(Perfect Simple Table.Level1 :Serial Number 1)

Afamatrix HT.+3. L2:7
(Hyper Table.+3 Factor.Level2.:Serial Number 7)

Afamatrix PHT.+3. L2:7
(Perfect Hyper Table.+3 Factor.Level2.:Serial Number 7)

Afamatrix IST.L3:5
(Imperfect Simple Table.Level3.:Serial Number 3)

AfamatrixIHT.*7. L6:90
(Imperfect Hyper Table.*7 Factor.Level6:Serial Number 90)

	Variation Code		
	Level Number		**Computation Serial Number**
Afamatrix	1	:	1
Afamatrix	2	:	56
Afamatrix	3	:	24

Afamatrix Puzzle Combination Code

The Afamatrix Combination Code also makes the Construction of Afamatrix Magic Square Grid Puzzles easy. This is because the codes contain information on the Sub-Level and Serial Number of Selected Afamatrix Magic Square Grid Puzzle. Just looking at the code, one can easily know the required Magic Square Grid Puzzle which can only be selected from the specified Table.

Therefore to construct a unique Afamatrix Magic Square Grid Puzzle, one will need a combination of the Afamatrix Magic square Variation Code and the Afamatrix Magic Square Conjugation Code.

Examples of Afamatrix Combination Codes

AfamatrixSL2:89
(Difficulty Sub-Level 2:Serial Number 89)

	Combination Code		
	Sub-Level Number		**Combination Serial Number**
Afamatrix	1	:	6
Afamatrix	5	:	55
Afamatrix	3	:	14

The notations can simply be referred to as Afamatrix Code, When both the Magic Square Grid Variation Code and the Magic Square Grid Puzzle Combination Code are written together. Examples are shown below.

Examples of Afamatrix Codes

Afamatrix ST.L1:1-SL2:89
(Simple Table.Level1.:Serial Number 1-Difficulty Sub-Level 2:Serial Number 89)
Afamatrix PST.L1:1
(Perfect Simple Table.Level1.:Serial Number 1)

Afamatrix HT.+3. L2:7
(Hyper Table.+3 Factor.Level2.:Serial Number 7)

Afamatrix PHT.+3. L2:7
(Perfect Hyper Table.+3 Factor.Level2.:Serial Number 7)
Afamatrix IST.L3:5
(Imperfect Simple Table.Level3.:Serial Number 3)

AfamatrixIHT.*7. L6:90
(Imperfect Hyper Table.*7 Factor.Level6:Serial Number 90)

Table 1

	Variation Code				**Combination Code**		
	Level Number		**Computation Serial Number**		**Sub-Level Number**		**Combination Serial Number**
Afamatrix	1	:	1	-	1	:	6
Afamatrix	2	:	56	-	5	:	55
Afamatrix	3	:	24	-	3	:	14

Some Examples of Afamatrix Magic Square Grid Puzzles

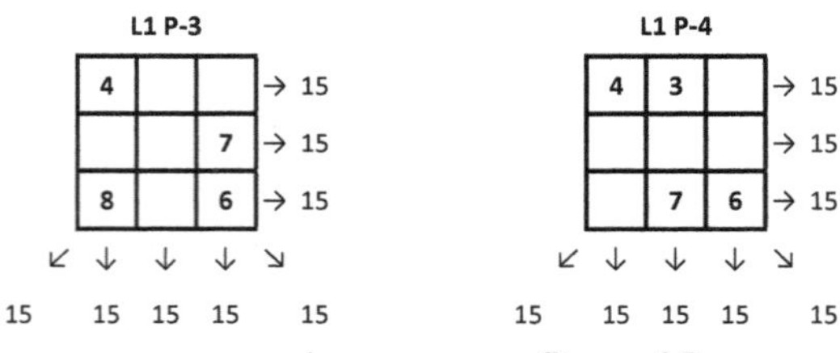

Answers on Page 45

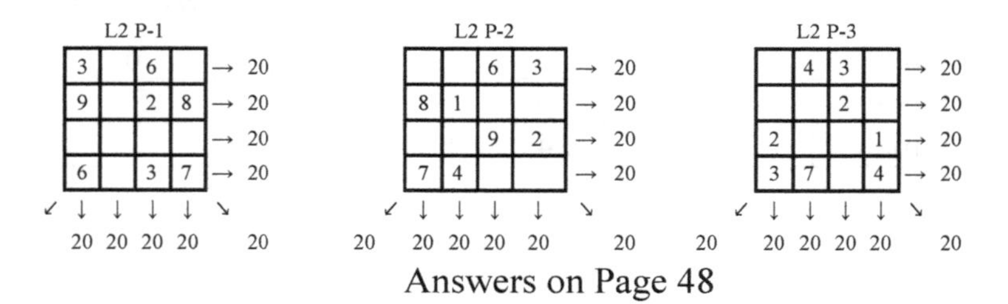

Answers on Page 48

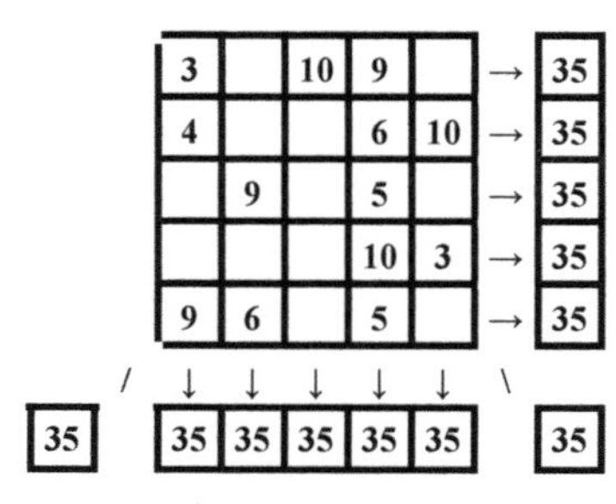

Answers on Page 57

↗ 90

2		4	2		6	4		2	6		2	6	1	8	8	3	4	→ 90
7	5			5	1			7	1		9	7					9	→ 90
	1	8	4		8	8	1	6	8		4	2	9	4	6	7	2	→ 90
		4	2	7	6	4		2	6		2	6	1	8	8	3	4	→ 90
7	5	3			1		5	7	1		9	7	5	3	1	5	9	→ 90
		8	4	3	8	8	1	6									2	→ 90
2		4		7	6			2	6		2	6	1	8	8	3	4	→ 90
		3	9		1	3	5	7	1		9	7	5	3	1	5	9	→ 90
6	1	8		3	8	8	1	6	8		4	2	9	4	6	7	2	→ 90
2		4	2		6			2	6		2	6	1	8	8	3	4	→ 90
	5	3	9	5	1	3	5	7	1			7					9	→ 90
6		8			8		1	6	8		4	2	9	4	6	7	2	→ 90
2	9	4		7	6	4		2	6		2	6		8	8	3	4	→ 90
7	5		9	5	1	3	5	7			9	7		3			9	→ 90
		8	4		8		1	6	8		4	2		4	6	7	2	→ 90
		4	2		6			2	6		2	6		8	8	3	4	→ 90
7				5	1	3		7	1		9	7		3			9	→ 90
6	1		4	3	8	8	1	6	8		4	2	9	4	6	7	2	→ 90
↓ 90	↓ 90	↓ 90	↓ 90	↓ 90	↓ 90	↓ 90	↓ 90	↓ 90	↓ 90	↓ 90	↓ 90	↓ 90	↓ 90	↓ 90	↓ 90	↓ 90	↓ 90	↘ 90

Answer on Page 53

References

1) Chapter 2 –Understanding Magic Shapes
2) Chapter 3 –What is a Magic Square Grid?
www.wikipedia.com/magic-square

3) Chapter 3 –History of Magic Square Grids

a. ***www.wikipedia.com/magic-square***

b. ***Lo Shu – Magic Squares (http://www.penninetaichi.co.uk/index_files/Page351.htm)***

c. ***www.textilemuseum.ca***

d. ***https://www.eventbrite.ca/e/magic-squares-artists-panel-the-manifest-and-the-hidden-tickets-1360271609***

e. ***http://www.math.buffalo.edu/mad/Ancient-Africa/mad_nigeria_pre-colonial.html***

Other Publications by the same Author.

1) Afamatrix Puzzle Game Series
2) Just Puzzle and Fun Family Magazine
3) The Grand Finale
4) Religious Tolerance and Harmony, the Yoruba Example.
5) Essential Compilations

Contact Information

For further enquiries, Subscription, and Distribution, Contact:

Akalusia Ent,
House 27. Road 19,
Upper-North,
Trans-Ekulu, Enugu, Nigeria

afam.aniagu@yahoo.com

+234- 8030887172

CPSIA information can be obtained
at www.ICGtesting.com
Printed in the USA
LVHW022345030423
743403LV00015B/1586